BLOOD AND SOIL

Richard Walther Darré and Hitler's 'Green Party'

THE KENSAL PRESS

British Library Cataloguing in Publication Data

Bramwell, Anna
    Blood and soil: Walther Darré and Hitler's
    Green Party.
    1. Darré, Walther   2. Politicians—
    Germany—Biography
    I. Title
    943.086'092'4   DD247.D3

    ISBN 0-946041-33-4

Published by The Kensal Press
Kensal House, Abbotsbrook, Bourne End, Buckinghamshire.

*Typeset, printed and bound in Great Britain by Butler & Tanner Ltd,
Frome, Somerset.*

'Some of his ideas were novel and somewhat bizarre, but it is not a crime to evolve and advocate new or even unsound social and economic theories'.

—Judgement on Darré, *TWCN*, xvi, p. 555

'En effet, la haine est une liqueur précieuse, un poison plus cher que celui des Borgia—car il est fait avec notre sang, notre santé, notre sommeil, et les deux tiers de notre amour! Il faut en être avare!

—Baudelaire, 'Conseils aux Jeunes Littérateurs', *L'Art Romantique*, p. 58.

Richard Walther Darré in (*F. Krausse, Goslar*)

Preface

This book is a political biography of R. Walther Darré, Minister of Agriculture under the Nazis from 1933 to 1942, and populariser of the slogan 'Blood and Soil'. As the title of the book implies, he headed a group of agrarian radicals who, among other issues, were concerned with what we now call ecological problems. While not a formal party grouping, they were a significant enough power bloc to perturb Hitler, Himmler and Heydrich. Although it was not possible to provide a comprehensive biographical study of the Green Nazis in the space available here, it is hoped that this biography of Darré will be a step in that direction.

I am grateful to the staff of the City Archive, Goslar, the Federal Archive, Coblenz, the Institute for Contemporary History, Munich, and Mr Wells of the Wiener Library, London, for their valuable assistance and courtesy. A grant from the British Academy enabled me to complete my research in the Goslar and Munich Archives. Correspondents are too numerous to list here, but an exception must be made for the Registrar of King's College School, Wimbledon, who kindly answered my questions about their old boy, Darré, while special thanks are due to those who opened their personal files and their memories to me, and who, across a gap of language, generation, culture and conviction, tried to communicate their perspectives. Ursula Backe permitted me to read and use the restricted Backe papers at Coblenz, and use extracts from her diary, while Frau Ohlendorf, Frau Meyer, Hans Deetjens, F. Krausse, Professor Haushofer, Dr Hans Merkl, Dr R. Proksch and Princess Marie Reuss zur Lippe granted interviews and wrote at length answering my questions.

I would like to thank Mrs Celia Clarke, Press Officer of the British Warm-Blood Society for information about the Trakhener horse, and I am greatly indebted to Dr John Clarke, Reader in History, University of Buckingham: Fellow of All Souls' College, Oxford, for his long-term interest in and encouragement of this project.

For permission to use material that first appeared in my article 'R.W. Darré: Was He Father of the Greens?' (September, 1984) thanks are due to Juliet Gardner, editor of *History Today*, also to the editors of the *Journal of the Anthropological Society of Oxford* (*JASO*), for their kind permission to reproduce extracts from my article 'German Identity Transformed', published in *JASO*, vol. XVI, no. 1, Hilary, 1985.

Historians are becoming more and more modest. It is now customary to regard the writing of a history book as essentially a collective act, where praise is due to the collective body, and blame alone accrues to the author. Writing this book was an unfashionably individual and solitary activity, and praise as well as blame is mine alone. I would like to stress, with more sincerity than is usual, my sole and complete responsibility for the contents.

Contents

List of Illustrations

Richard Walther Darré. (*F. Krausse, Goslar*)   *Frontispiece*

Between pages 152 and 153

Introduction

In 1930, European agriculture faced one of its worst depressions. Peasant movements, farmers' leagues and agitation for land reform became politically active in many countries. In some, the rhetoric had left-wing overtones: in others, it carried a vehemently radical nationalism. Generally, it was populist and anti-institutionalist, and potentially violent.

In one European nation, a government came to power partly through the votes of disaffected small farmers. It proceeded to introduce laws establishing hereditary tenure for small and medium sized farms. The wholesale food industry was virtually abolished, and a marketing system established which set prices and controlled quality. Later, quotas were introduced. This corporation was run by a quasi-independent 'quango'. Some two thirds of its farmland was taken out of the free market by the laws, and control of a farm which came within the regulations was made conditional on farming ability. A Back-to-the-Land programme was introduced, which established viable peasant settlements, and poured money into the rural infrastructure where the settlements were located. A drive to increase peasant productivity was introduced, which was remarkably successful in coaxing more productivity per hectare from the land, and in increasing intensive agriculture. The agricultural experiment lasted some six years, until the country went to war. While the legislation remained on the statute book, agriculture fell under wartime controls, and the Back-to-the-Land movement tailed off.[1]

In the 1930s, the name of the man responsible for this legislation, this experiment in anti-capitalist agriculture, was widely familiar.[2] Today, it is virtually unknown, except where academic text-book writers have mentioned him in passing as a trivial, insincere fool. The programme outlined above has been labelled with adjectives like 'fanatical, irrational, insane, primitive, inhuman, monomaniac, brutal, bloody, reactionary, unproven', and even 'unprogressive', a particularly crushing blow. The legislation itself has been described as unreal, Utopian, wishful thinking, not particularly original anyway, and insincere. Primitive and mystical are probably the most commonly used adjectives, with insane running a close third.[3]

But why is this? Why should measures which have been advocated by people of all political persuasions as solutions to problems of land tenure and agricultural marketing in the Third World today be considered primitive and insane? The answer is, that the country in question was Nazi Germany. This has affected the history of its agricultural experiments in several ways. First, historians who specialise in one country tend in any case to see that country's problems and policies as unique. This applies with much greater force to historians of Nazi Germany. There has been a reluctance to examine these policies in comparison with those of the Successor States in the 1920s and 1930s, to say nothing of the problems of peasant societies today. The existence of a large (in comparison with Britain) agricultural population—some 29% in 1930—has been seen as something to do with Germany's lamentable lack of western European humanitarian democracy, part of a structural political problem which prevented Germany, as Golo Mann says, fulfilling its true role in the world, that of being the first nation to combine advanced industrialisation with truly humanitarian, internationalist, social democracy.[4] Reading history backwards has its problems, especially when it is done from the highly politicised (and nearly always social democratic) viewpoint natural to historians of Nazi Germany. The second problem is that the agricultural sector tends to be seen exclusively in its function of non-urban, hence non-urban liberal/democratic, obstruction to progress. German landowners, despite their homely, impoverished lifestyle, are associated with Prussian militarism and the German élite that ruled through the caste-oriented political structure of the nineteenth century, and never have had a good press, then or now (nor do they receive much sympathy in this work, an omission which the author hopes to rectify in a future book).[5] Historians, then, who on the whole realise that they are an urban luxury, see no reason to sympathise with a sector which is associated with incest-in-the-pig-sty at the lower socio-economic end of the scale, and beastly Junkers at the other. Third, National Socialist Germany is seen loosely (if mistakenly) by many historians as an extension of right-wing, neo-conservative, capitalist, authoritarian and élitist policies, coupled with an exclusivist, racialist nationalism carried to the point of self-destruction. Historians who feel in themselves a warm sympathy towards neo-conservative, capitalist, authoritarian or élitist policies will shudder and go away, muttering as they go that Stalin was just as bad, anyway, but leaving the field open to internationalist social democrat sympathisers, who can only examine the exceedingly broad church manifestations of National Socialism (from technology, to art, architecture and science, for example) through their own progressive prejudices. Not unnaturally, they soon run out of categorisational capacity. Some recent work, scholarly and scrupulously annotated as it is, seems to consist largely of desperately strung together adjectives. National Socialism is romantic, and modernistic: it is technocratic and backward-looking. It is petty bourgeois and has aristocratic pretensions: it is reactionary and pragmatic. The frenetic air that characterises so much recent writing on the period—carried out, one must stress, by fine scholars—shows that an obstruction has been reached.[6] How can one summarise simply and clearly a movement that moved from being anti-parliamentary and revolutionary to a mass, legal parliamentary movement: that had only six years of peace and five years of war; that rapidly ran the gamut from populist radicalism to dictatorship: that seems unique, yet shows continuity. How is one to label those Germans who continued to work under National Socialism, in the field of scholarship, science and, of course, everyday life? Are all their products, their achievements, their errors, too, to be considered as archetypal National Socialism? Might they have developed that way in any case? How can one distinguish what was German in 1933–45 and what was Nazi?[7]

The need so to distinguish is demonstrated by the tangle of conflicting categories that historians have created, so that on the whole one has to turn from political and diplomatic historians for a comparative and contextual examination of this period, to historical geographers, economic historians, historians who specialise in science, or architecture, and military historians.[8]

In many other areas of intellectual history, emphasis has shifted over the years from one aspect of a movement to another. Sometimes there is a consensus over what is the 'real' phenomenon. For example, Newton's gravity theory is seen as the real Newton, while his Rosicrucianism is seen as a kind of irrelevant extra. Marx, hardly taken seriously by academic economists until Joan Robinson revived him in the early 1940s, is now widely seen as an important theoretical economist in his own right, the last and greatest classical economist. But there is a sturdy and growing industry of those who feel that the real importance of Marx lies in his role as precursor of Hitler and all the ills of the twentieth century: racialism, anti-semitism, genocide and Social Darwinism. This interpretation of the 'real' Marx may overtake and hide the Joan Robinson interpretation.[9]

With Nazi Germany, it is not difficult to see its war-time policies as the 'real' Nazism, and everything connected with it up to 1939, and by extension up to 1933, as leading to that point. This interpretation has been strengthened by its role in maintaining the validity and necessity of immediately post-war Allied policies, and, indeed, to justify the paradox that we fought a war to save democracy and virtue in alliance with a power, Soviet Russia, which had already by 1939 carried out more murders, probably some twenty million, than Nazi Germany was ever to be responsible for, and that in peace-time. Many historians of the period had themselves played a part in shaping policy during and after the war: some had worked in British Intelligence. But the link between pre-Nazi and early Nazi ideas and later Nazi practice remains a problem; it is hoped that this book, by examining Darré's agrarian and eugenic policies, in concept and in practice, will contribute to establishing the simplest of historical data, what actually happened: *wie es eigentlich gewesen ist*.

WHY A POLITICAL BIOGRAPHY?

History is made up of men and men's actions. 'Men make their own history,' and in those rare epochs where that history is created on the basis of an idea it can be much easier to establish the clarity of an idea than the clarity of factual data. Behind the historian's generalisations lies the ever-present awareness that 'under each grave-stone a world lies buried'; that the texture of reality is composed by a myriad of individualities.[10]

German historians tend towards a theoretical approach. National Socialism, especially, has attracted the apparatus of conceptual frameworks and sociological categories. Categorisation is an aggressive act, and sociology is an aggressive science, even more than history. It is hard to deal with individuals in this harsh and impersonal fashion, which is why sociologists categorise people in groups. In the case of Richard Walther Darré, Minister of Agriculture and Peasant Leader from 1933 to 1942, we are dealing with a man who overturned many pre-existing theories, and who seemed to have performed the historically unacceptable. He affected the course of history by having the particular ideas and abilities that he had, at the right place and at the right time. A large tranche of National Socialism's voting support in the crucial election of 1932 came from the small farmers of the north and north-west. The influential farmers' leagues and *Bauernvereine* were taken over by cadres organised and led by Darré. He was the author of two best-selling works in the late 1920s which propounded the virtues of the peasant and the need to re-organise society through rural corporations.[11] In short, it seems worth while to trace the life and career of one man, in order to see what deductions can be made from this particular case to the general.

This book is the political and intellectual history of a man described in highly coloured terms by admirers still living today as the 'last peasant leader'. It traces his ideas, his decision to turn to politics, the continuity of his views till his death in 1953, his successes and his vast, painful failure. It has wider significance because Nazi Germany was the one country in western Europe where radical agricultural policies were actually introduced, not just demanded. Darré becomes an important figure, then, in a way that other early Nazis are not. Feder's Social Credit died without issue, unless one were to count Keynes, who was a fan of Silvio Gesell, one of Feder's inspirers.[12] Gregor Strasser, that very Prussian mixture of authoritarian socialism and social élitism, lost out to the south German influence by 1932. Feder and Strasser did not see their ideas carried into effect: Darré did. That alone makes him significant in the world of revolutionaries. As was said of Bakunin, such men are necessary in the first days of the revolution, then they have to be shot. In its slow motion essence, that is what happened to Darré. Unexpected successes emerged from his legislation. Productivity on small farms rose: they became more efficient and more modernised. The demoralised farming population of Germany, especially the economically innovative but socially conservative north-west German farmers, recovered their self-image and confidence. But conditions remained relatively worse than in towns, and the drift from the land continued. Farms could not be seized by banks, but farms seized by banks before 1933 were not returned to their owners, and the improvement in agricultural conditions meant that farms were in any case less likely to be foreclosed upon. Above all, urbanisation continued. Autobahns and the many technical achievements of the 1930s stand out in our impressions of the social history of the period. The most significant conclusion to be drawn from a study of the results of Darré's policies is that they failed to have the dramatic results expected of them, and that is a conclusion that has serious implications for planners trying to aid peasant societies today.

In the discussion over what is and is not National Socialism, agricultural economics is of special interest, as it is an area where there is a clear continuity of personnel and ideas from before the First World War to well after the Second. There are only so many ways in which one can evolve solutions to the problems of small peasant farmers, and if one wants to maintain them as a class, the solutions available shrink dramatically to the inalienability of land and the maintenance of prices. This continuity of ideas has perhaps been obscured by the dramatic propaganda value of the phrase 'Blood and Soil'. This slogan, coined in the early 1920s by a renegade Social Democrat, became virtually identified with National Socialist ideology, to the extent that a recent work on Nazi cinema actually called its chapter on ideology, 'Blood and Soil'.[13] The element in this slogan that concerns the link between people and land has been either neglected or jeered at uncomprehendingly, as recorded above. Yet in many countries today, including European nations such as Greece and France, and several States in the United States of America, farm purchase by non-nationals is either forbidden or tangled up with so many booby traps as to be made extremely difficult. The position in the Third World is of course much more exclusivist and racialist. Sometimes the rhetoric of nation and race and land is open, as with Israeli settlers in Arab lands in recent years.[14] Then the urban liberal may feel a frisson of discomfort. But usually it is implicit rather than explicit. Peasants like to keep their land. They like to live among other peasants, preferably near enough to a town where their produce can be marketed, but far away enough to avoid pollution by the non-kin. These qualities are known to us: they are part of our mental furniture, part of what we mean when we say 'peasant'. We think it is rather cute. People have a right to their land, to their peasantness. Sensitive observers even understand the gut dissatisfaction felt by all food producers when they exchange their produce for money. Market exchanges satisfy only when value is exchanged for value; and no-one who has grown his own produce can feel satisfied with the exchange of his goods for money; because the produce he sells is the result of life, it contains life, and what he receives is dead. There can be no equivalent. Hence the conundrum that no one is more money-grubbing than the peasant, yet no one more contemptuous of it.

Yet peasants have another quality; they may crave continuity, but they are also innovative. The agricultural surpluses produced by efficient peasant technology are thought to be the basis for successful industrialisation. This means that given the right marketing structures, it should be possible for efficient peasant production to survive urbanisation and co-exist with it. One agrarian historian has called this type of producer the 'smallholder technician of the future'.[15] The populist, anarchist, political impulses associated with the European peasant may mean that he will not adapt to the urban hierarchy inherent in democracy; but that is not to say that he cannot survive literacy and television. The Russian agrarian economist, Chayanov, wrote a science-fiction novel in the dark days of 1921 in which he envisaged prosperous peasants (kulaks, perhaps) attending concerts in towns in their own aeroplanes, then returning to milk the cows. This vision of the independent, prosperous, free smallholder, so prominent in German agrarianism, is curiously similar to the English ideal of the yeoman farmer. That great neo-Englishman, Jefferson, wanted to see an America composed of independent gentleman farmers. Jefferson was elected an honorary member of the Bavarian *Bauernvereine* (Peasant League), of 1810. The American West provided inspiration to a German geographer and economist, von Thünen, who envisaged an economy based on farming, with large estates divided up among labourers, who would be able to accumulate enough capital to buy their land after ten years work. Von Thünen, widely attacked by Marxist economists today as the founder of the idea of marginalism, left his land to be divided up among his own labourers, and is claimed by East Germany today as an early Socialist. He was an inspirer of Herbert Backe, fervent National Socialist and Darré's successor as Minister of Agriculture, and also of Chayanov, the Soviet agronomist.[16] The point of these apparently disconnected observations is that accepted political and social axes break down when ideas about peasant society and production are concerned. The small farmer, the peasant, is not just a vehicle for the townsman's romantic or reactionary ideas, he can be seen as the core of society itself, the most productive, the most resourceful, the most innovative group of a nation.

At a time when accepted orders, ideas and structures were widely perceived to be collapsing, the peasant in Germany offered not a return to some golden age, but a sound, healthy starting-point to something new. *Reculer pour mieux sauter*; to get rid of these existing constraints, seemed alluring to many neo-conservative intellectuals of the 1920s. German agricultural economics was traditionally a *Volkswissenschaft*: it had a social and sociological dimension. Agrarian ideas belonged to the pre-neo-classical tradition. The word 'peasant' in Germany did not have the connotation of serfdom, or previous bondage to a landowner, that it has in English. To give one example of the small but tangible linguistic difference involved, hunting over peasant land with horses was forbidden in most German states in 1848; an example of the relative self-confidence enjoyed by the German smallholder. Part of this confidence was reflected in the extent to which the small farmer was seen as the core of future agricultural productivity, and indeed, peasant farming showed an increase in both productivity (higher for small farms than for large farms from 1860 to 1925) and in the number of small farms, even at a time of acute agricultural depression.

Nazi rural ideology was a fusion of several different streams of ideas and benefited from the strength of existing rural traditions. There was an anti-Prussian and anti-state element in peasant political behaviour. Small peasant producers were hostile to perceived market inadequacies and to credit constraints. The Land Reform ideals of Henry George found a home in German Land Reform movements of the late nineteenth-century; they differed from Darré's plans in proposing a quasi-nationalisation of land through a land tax, but were similar in the basic ideal of the independent small farmer. Corporatist, anti-liberal economists found it easy to understand the idea of the peasantry as a separate *Stand*, as did the rather more liberal economics associated with Nordicist circles (who believed in the inherently freedom-loving nature of the northern Europeans).

Peasant production was in fact seen as materially superior as well as morally superior. The peasant would work harder, know his land better, be more economical in bought-in inputs, hand-dig his land, and coax more from the soil. The idea that peasant farming could be *economically* desirable, and lessen dependence on imported fodder and food, naturally gave impetus to the *moral* arguments—that the peasant represented 'freedom, property thrift': 'frugality, loyalty, hard work'.[17] The desirable norm was the self-sufficient family farm of fifteen to seventy-five hectares; not too small, not too big. Many of the dignified, white-haired professors whose portraits can be seen adorning text-books on agricultural economics became attached to Darré's entourage: they translated the words 'Blood and Soil' into a text-book for agrarian reform, for renovation of silo buildings, a good example of the way in which ideas change and reform as they sift through men's minds.

Indeed, the typical agricultural economist who supported Darré and the Nazis in the 1930s would have been born in 1895, son of a village schoolmaster in a small village. He would be educated in a republican and atheistic tradition, but would acquire a theistic, but probably anti-confessional religious feeling in later life. He would write a doctoral dissertation on peasant productivity, work as adviser to a farmer's union, and become a professor some time in the 1920s. He would visit Soviet Russia to help negotiate seed or horse-breeding stock purchases, and be roped in to help the infant National Food Estate after 1933. He would admire Darré's support for the peasantry, and his attacks on the large landowners and 'Prussianism', but consider his tendency to write off all peasants south of Göttingen a mistake. He would weary of the slogan 'Blood and Soil' without relinquishing support for what he perceived Darréism to be about, be drafted into the SS advisory service in 1939, sent to Russia to improve milk production, write memos dissenting from German treatment of the Ukrainians in 1941, and be in and out of concentration camps until 1945. Then he would, after de-Nazification, take up a post again as an agrarian economist or historian or civil servant, and go on writing articles about the importance of the peasantry to the body politic, and praising Darré's support for methane gas plants and small tractors. He would support the EEC Agricultural Policy with enthusiasm. His writings would have a curiously unanalytical and anodyne flavour that would puzzle and irritate later writers in *Peasant Studies*; caused by his inability to expound on his *reasons* for supporting the peasantry in post-war Germany. He would be utterly perplexed by liberal, free-market economics, and believe that England's fields had grown only thistles since the Reform Act. The sight of the farmers of Schleswig-Holstein marching on Hamburg in the company of anti-nuclear Greens would stir a faint echo of memory, a sense of *déjà vu*. To a man who had helped pass a law in 1934 ensuring that new woodland plantings were composed of mixed deciduous and evergreen trees, who had seen nature reserves established, and assumed that afforested land was sacrosanct (not only was it not planned to cut down trees for peasant settlement but landowners offered to swop their arable land for publicly owned forest so that the public authorities could settle the arable land), an opinion poll that showed that 99% of Germans had heard of the dying forests, and that 74% of them showed great concern, would have caused no surprise. In a sense, this work should be dedicated to this man.

At the beginning of this introduction, I stated my premise, that Germany, in its 1930s agricultural policies, was the one European country that instituted radical tenure and price legislation in an attempt to solve the problems of peasant producers. Why, then, is it so often argued that German agrarianism was the product of a romantic, reactionary, fanatical reaction against industrialisation? After all, between the wars, Poland, Romania, Bulgaria, France, Italy, Spain and Denmark had peasant parties and agitation for land reform, and, in the case of the Successor States, land reform did take place, albeit at the cost of an ethnically alien landowning class.[18] Britain also had its smallholdings movement, and in the Edwardian era, its land reform movement. The ideal of a kind of yeoman farmer as the basis of a nation was widespread in British colonial plans in the 1930s. These countries have nothing in common; they display quite different rates of industrialisation, ranging from early and complete, to late and impressive, to non-existent. German fears about the disappearance of the peasantry date back to Frederick the Great, while the diminution of the tax base in Oldenburg in the 1820s, caused by mass peasant emigration, incited the creation of a settlement commission. Was Frederick the Great reacting against industrialisation? Was Freiherr vom Stein? One begins to wonder why it was that Germany went on modernising so very slowly, over so many decades, and reacting so much.[19] The value of comparative history is that it allows these easy clichés to be tested empirically. However, one fears that it will be some time before the work of economic and agricultural historians like Farquharson and Warriner creeps into the straightforward political histories, and until that time the German desire to have a populated, prosperous countryside, farmed with tender loving care, will be seen as 'fanatical', simply because it was a belief held by men who found it possible to work with the National Socialist government in its early years.

But if agrarianism was not backward-looking and primitive, what did it seek? Can one integrate 'naturism' with the 'smallholding technician' ideal? One answer is that Darré and his circle laid stress on the innovative aspect of the small farm, and supported methane gas plants, small tractors suitable for small farms, and market garden machinery. When a British team of agricultural experts visited Germany after the war in search of new inventions, they were surprised to find developments in dwarf-rooting stocks for fruit trees, roto-tiller and spraying equipment, bulk manures based on peat and dung, and lightweight machinery. In some areas, such as fungicides and seed-cleaning machinery, the Germans were more advanced. These were all innovations designed to help the small farmer, and their success implies that if there had not been a war, Nazi Germany would have continued to develop its agriculture along these lines.[20] It was not really as paradoxical as it seemed that German agriculture should be developing in this way in an era of technological change. Innovation implies the capacity to escape from old methods, the dead-weight of large farms, large companies, banks, outworn institutions, and all the encumbranced interest groups fighting over the economic corpse of the 1920s. Nature-based thinking tends to optimism, rather than to appeals to the past. Its essence is to be 'forward-looking', because of its inherent rejection of the old, the traditional. It emphasises youth, the young, the new, and this fits in with the demographic pattern of Nazi support in the late 1920s. When one agronomist exhorted his farmers to go 'Forward, onward, avoiding the old errors', this was an archetypal National Socialist formulation: its similarity to the progressivist 'onward ever upward' slogans of twentieth-century Britain is not accidental, because both were the product of a radical rejection of existing structures.

The word 'radical' is not used lightly here. The naturist thinker is always Antigone, not Creon. Nature is seen as a path which leads somewhere; it is a teacher. One goes to the natural world to learn from it, and returns with a series of lessons. Nature teaches that there *is* a truthful, real world, which can, though with difficulty, be seized, grasped and verified. It exists objectively. Why is this apparently obvious attitude a radical one? Because conservative thought is either indifferent to this sort of realism—preferring criteria of social usefulness—or else translates reality to a metaphysical plane where it poses no threat to social stability. Socialists and communists believe in structures and, once in power, in stability. At heart, they do not want to rock the boat, they want to get in it.

But the man who goes to nature for his beliefs is rejecting these compromises. He may be of an unanalytical cast of mind, but he knows how to say no. He is inherently suspicious and bloody-minded. He suspects tradition, ruling classes, and lies, even holy lies. He prefers kin to caste. He can not, I think, be described as Utopian or mystical, just because he does not conform. If one defines Utopianism as the attempt to escape from 'the Wheel', then nature-inspired reformers are not Utopian.[21] They go to nature to learn, and return with the recommendation that one clings to the Wheel, because it is the most sensible path of action.

Whether or not Darré's premise that the northern European peasant was the most valuable and creative element in European civilisation was true, is unprovable. It could be argued that any act of political self-identification is irrational. Given his premise, Darré's method of argument and system of ideas was rational. His attempt to fulfil his political ideas through political action rested on the assumption that the institutions of a revolutionary state could be used to achieve revolutionary aims. It is this assumption that now appears irrational.

Certainly, Darré's ideology was something more than a purely teleological policy, though; something which is not completely amenable to analysis in terms of polycracy, productivity or eugenics. In reacting against what he saw as the failed, finished institutions of his day, the cultural colonisation by western Europe and even America, Darré was implying a cultural criticism that he was not competent clearly to express, only to label. When Heidegger talks of the consumption mentality as 'the organisation of a lack', which exists only to fulfil the vacuum of 'non-Being', when he talks of 'the world which has become an unworld ... the desolation of the earth', he was vividly expressing a sense of alienation, of being lost in a vacuum of unreality, that lay behind Darré's thinking.[22] Darré was to write before his death that he had been a fool to think that the Nazis could have repaired the broken link between man and soil, nature and God. But he continued to think that there was a tangible, real world, veiled by man's error, which could be reached, if only the veil could be torn aside.

> The unnoticeable law of the earth preserves the earth ... in the allotted sphere of the possible which everything follows and yet nothing knows ... The birch tree never oversteps its possibility. The colony of bees dwells in its possibility ... It is one thing just to use the earth, another to receive the blessing of the earth ... in order to shepherd the mystery of Being and watch over the inviolability of the possible.[23]

It is the core of my argument that one should not let the existence of the uniforms and the swastikas interfere with the evaluation of Darré's attempt to 'watch over the inviolability of the possible'. He was guardian of a radical, centrist, republican critique which pre-dated National Socialism, and still lives on. That the political development with which he and others associated themselves reversed and then destroyed that viewpoint is an irony that will echo and re-echo many times in this work.

CHAPTER ONE

Who was Darré?

1896-1924

Like many National Socialist leaders Darré came from outside Germany. Hess, Backe, Rosenberg, and, of course, Hitler, were all born outside the frontiers of the German *Reich*. 'Auslanddeutsche' were prominent and active in German nationalist circles, whether because of the shock of returning disappointed to a longed-for homeland, or the outsider's clearer perception of national ills. Not only was Darré born in Argentina, on 14th July 1895, but he did not see Germany until he was ten years old, when he travelled there alone to go to school. He was the son of a director of the trading firm Engelbert Hardt & Co., Richard Oskar Darré (who had emigrated to Argentina in 1876), and a half-Swedish, half-German mother, Eleanor Lagergren, daughter of a trader, who met Richard Oskar Darré in Buenos Aires. On his father's side, Richard Walther Darré was descended from a Huguenot family which had emigrated from northern France in 1680. His maternal grandmother was the daughter of a yeoman farmer in Lower Saxony. His maternal grandfather was a Swedish yeoman farmer, and his uncle was *Bürgermeister* of Stockholm. There were four children of the marriage: Richard-Walther, the eldest son, Carmen, Erich and Ilse. The home near Buenos Aires was in the prosperous suburb of Belgrano, a stuccoed three-story house with a formal front garden, surrounded by trees, and a cobbled lane outside.[1]

Darré's father, born in Berlin in 1854, was a typical member of the Berlin mercantile bourgeoisie, prosperous, educated and politically liberal. Although nineteenth-century Prussia is associated with cartels, étatism and militarism, this was also an era when German overseas traders and wealthy merchants were sympathetic to English constitutionalism and laissez-faire ideals. He admired the American educational reformer and feminist Ellen Keys, and in his memoirs, published in Wiesbaden in 1925, criticised the Prussian reaction to the revolutions of 1848. He had considerable contact with the English business and farming community in Argentina and Brazil, and many English friends. He considered a knowledge of the English language and culture essential for any serious businessman.

Before going to South America, Darré's father began to train as an engineer, but switched to medicine half-way through his studies. Before he had finished, the family was financially ruined, and he had to leave university. The President of the Berlin Stock Exchange, Richard von Hardt, an old friend of the family, stepped in, and he was put through a commercial training course and sent to Brazil. He spent several years in Brazil and Argentina. In 1888, he became a partner of Engelbert Hardt & Co. in Buenos Aires.[2]

Having been forced to abandon the subject on which his heart was set, Darré's father became a great believer in discipline and renunciation. He brought up his large family in this spirit, quite unsuccessfully, since his emphasis on the gloom and misery of life was rejected by them in favour of a somewhat Utopian hedonism. This rejection of the father's sternness was partly because, as Richard Walther observed on hearing of his father's death, Richard Oskar Darré had not behaved 'with that German virtue and honour he preached', but drank too much, and was a womaniser. He was rumoured to have an illegitimate son, supposed to be half-Scottish, who was brought up with the rest of the family.[3]

Darré attended the local German school until he was 9 years old. He then returned to Germany, at first alone, 'to be brought up as German'. He lived with Professor Elisabeth Gass at Heidelberg *en pension*, and finished his primary education at the local *Volksschule*, or primary school. In the autumn of 1905, he began in the *Oberrealschule* at Heidelberg, and remained there until 1910, when he was fifteen.

His school work at the *Deutsche Schule* in Belgrano had been adequate; concentration, behaviour and all subjects received good marks, especially in languages. However, once at Heidelberg, his achievement level plummeted. His French and English remained good, but his German and history came in for heavy criticism. For the first year he was in the bottom three of his class of about thirty-seven. After that period his marks fluctuated to such an extent that no pattern can be deduced. Where he excelled in one term he failed in the next. He seemed dreamy, well-behaved, and lacking in concentration. How much of this was due to differences in teaching methods between Argentina and Germany, or how much due to Darré's own attitudes, is hard to determine, but the change of eductional methods and schools may have affected his reaction to his schooling, since when he left Argentina his marks for French, Spanish and German, for behaviour and effort, were 'good'. while the school at Heidelberg complained of his lack of diligence and an 'unsuitable manner'. After five years his marks and class position were still low. At the Evangelical School at

Gummersbach, his marks quickly improved, with geography his best subject. Darré lost a year through his change of schools. His father was anxious that his son should learn something of the English language and way of life, and in 1911 he was sent as an exchange pupil to King's College School, Wimbledon. He had a gifted science teacher there, who seems to have had a considerable effect on him. He left King's convinced of the superiority of the English public school system to the more regimented German system, and with a fascination for English customs and political life. Another change followed in 1912, when he attended the *Oberrealschule* at Gummersbach for two years. His *Abitur* or university entrance examination was delayed by his changes of schools, and was now due to be taken in the Autumn of 1914. The war, of course, intervened.[4]

By the time he had left Gummersbach, his marks had improved, but it was clear that he was not suitable for an academic career, or for one of the professions. He was fond of open-air life. The vacations and the weekends were spent in excursions with his brothers and sisters, or with his school-friends. Outside school, however, he read widely and enthusiastically.

In Easter 1914, he attended the German Colonial School at Witzenhausen, south of Göttingen. Founded in 1898, in a large country house called Williamshof, in imitation of the English Colonial College and its training farms, it was designed to prepare future plantation administrators and farmers for the specialised conditions they would find abroad, and concentrated on practical farming, which was carried out on the 285 hectares of farm and market garden land owned by the school. Witzenhausen echoed Darré's earlier brief experience of English education at King's; the students were a self-governing body, carrying the aroma of the English prefect system. This system of self-government by student committees was held up by Darré in his later works as a model of education, because the system was less hierarchical than the normal Prussian one. Graduates of the school were awarded a diploma in colonial agriculture. It was an unlucky choice, since, after the war, this professional qualification was to prove inadequate in conditions of extreme unemployment, especially among the educated middle classes; while the loss of Germany's former colonies made the specialised nature of the training virtually useless.[5]

During this period, Darré felt torn between the idea of returning to South America to farm, or becoming a soldier. When war broke out in August 1914, he volunteered at once for the German Army, joining the Field Artillery Regiment no. 27 (Nassau) at Wiesbaden. He had spent one term only at Witzenhausen. Darré's 1933 biographer reports that of the hundred pupils who had enrolled in Easter 1914 with Darré, only ten would return after the war, and of the others, thirty were killed, and another thirty were too badly wounded to contemplate a farming career. The fact that he had volunteered remained a source of pride all his life: the point was that Darré was still of Argentinian nationality, and could easily have avoided the war.[6] He continued to have dual Argentinian–German nationality until 1933, when he became a minister; the Argentinian nationality then lapsed.

His wartime career was steady rather than spectacular. He served in two artillery regiments, both of which were virtually wiped out. He was wounded twice. In autumn 1916, in the Battle of the Somme, he received the Iron Cross, Second Class, and was sent on an artillery training course in January 1917. He was then promoted to reserve lieutenant. In July 1917, two shell splinters lodged in his left leg, and he was caught in a gas attack. He was then given a short home leave. He returned to the battles of Verdun and Champagne, and the last push by the German forces into France in March 1918. In April 1918, he was transferred to the Field Artillery Regiment von Scharnhorst, 1st Hannoverian no. 10, and fought with them until October 1918, when he contracted a fever, and was sent to a base hospital. Some weeks later Darré emerged to find that Germany had surrendered. His diary notes:

> One could see that politically speaking everything was in the air; but the total collapse of Germany was unexpected. The maddest rumours were everywhere. I myself was relieved from duty because of illness ... I looked for wounded comrades and in general completely disregarded politics.[7]

Darré and his fellow-soldiers suddenly heard the news from Kiel of the German naval mutiny and the election of a Worker's and Seaman's Council. On 7th and 8th November, 1918, his diary described how he saw 'Bolshevik agents' persuading soldiers to give up their arms and join the uprising. A soldier's council was elected at his barracks, Darré and two other officers remaining to negotiate terms with the representatives of the Social Democrats. Darré's family were living in French-occupied Wiesbaden. As soon as he had returned to his studies and was away from Wiesbaden, he joined a *Freikorps* regiment, the Hanover *Freiwilligen* (volunteer) unit.[8]

Darré's father was proud of his son's wartime record. The family had produced doctors and soldiers in equal numbers for several generations. He even published a compilation of Darré's letters home from the front, and of his war-time diaries.[9] However, Darré himself did not dwell on his wartime experiences, apart from writing articles on field artillery tactics in 1923 and '24. He joined the Stahlhelm, the German nationalist veteran's association, in 1922, while studying at Halle, and remained a paid-up member until 1927.[10] His health was probably permanently weakened by his war years. Between 1919 and 1923 he was under a course of homeopathic treatment for heart trouble and various nervous disorders. His regiment, the von Scharnhorst, had taken part in thirty battles between 1914 and 1918, and Darré's record shows that his personal participation in the war probably included as many engagements, in a fighting capacity.[11]

Although after the First World War Darré's father mourned 'all the inspiration, the joy, willingness of self-sacrifice, all hopes of victory, and happy future of the fatherland come to naught'; nonetheless, he retained at first much of his earlier liberal and pro-English spirit.[12] His attitude was more moderate than some of the typically civilian attitudes of the war years, expressed in such matters as a hostility to the English and French language, and attempts to replace foreign loan words with German root words. Fines were levied on those who used French, English, Italian or Russian words.[13] His attitude hardened after the First World War, when the family, now settled in Wiesbaden, found themselves in the French zone of occupation. Letters were censored, a pass was needed to move in or out of the zone, and currency restrictions were in force. Most of the family's substantial funds were lost in the inflation of 1923, while the increasing nationalism and militancy of the Argentinian government meant that problems arose remitting funds from Darré's old company in Buenos Aires. Darré's father wrote regularly to his son, and the two exchanged political analyses as frankly as they could, although occasionally the elder Darré would observe that censorship had interfered with a letter. In late 1923 they agreed that it would be safer not to discuss politics any more in their correspondence.[14]

The trauma of Germany's defeat in 1918 was followed by a series of defeats at the Conference of Versailles. By the time the Allied Powers had satisfied the territorial claims of the new states, more territory and resources had been taken from Germany than anyone had intended. Reparations were set at an unrealistically high level. Feelings in Germany ran highest in Upper Silesia, where a state of near-civil war existed between Germans and Poles, and in the Rhineland, where the French maintained an armed presence. The returning soldiers blamed the civilians, and especially the politicians at home for the defeat. This was partly because the defeat had been unexpected for many. The war in the east had apparently ended with victory. Under the Treaty of Brest-Litovsk, signed in March 1918, the new Bolshevik government had made peace with Germany, and had ceded considerable territory to her, including former Russian Poland, the Baltic States and the Ukraine. Massive settlements were planned in the Baltic lands for some 250,000 German farmers. After the Armistice of November 1918, the army considered continuing the war in the east, the Treaty of Brest-Litovsk having been revoked by Russia, but rejected it as impractical. Not only was this newly won territory given up, but land which had been under German rule and occupation for many centuries. East Prussia lost its hinterland; Danzig, a major trading post, was cut off from the rest of Germany. Many returning soldiers blamed the new Social Democrat government for the defeat, as much as or more than they blamed the revolutionary militants of 1918-19. 'We only lost the war, but the peace has been lost by others', was the accusation from a returning soldier in Erich Dwinger's war trilogy, *Wir Rufen Deutschland*. An ex-comrade from Darré's battery wrote to him in May 1920 that he had thought the reports of disaster and defeat in the French Press exaggerated, but after release from a French POW camp to Germany, 'we found how true it was'.[15]

The new democratic constitution signed at Weimar was unpopular with the Communists, with the conservatives, with broad middle-class opinion, and with the nationalist 'fatherland' movement. Furthermore, democracy was seen as alien to German political traditions. For the second time in 120 years, it had been brought in as a result of armed defeat, and was seen by many as being an essentially alien imposition; a *concomitant* of defeat. In the election, no one party gained a majority, and the Social Democrats had to govern in a coalition. The peace treaties had been signed only after painful consideration, especially the clauses accepting war guilt. Another potential source of opposition in the early 1920s lay in the ex-soldier himself. The army had at first refused to sign the treaties, and in late 1918 a league of ex-servicemen was formed, the *Stahlhelm Bund der Frontsoldaten*. At first a 'harmless group of old veterans', as early as 1919 it was active and vocal, and formed branches all over Germany, opposing both democracy and the 'System', as the new constitution was called. Its stance was conservative and nationalist.[16] The *Freikorps* was formed at the same time, consisting of ex-combat units and new volunteers, many of them too young to have fought in the war, and eager now to serve. It was originally called into being by the new Republic to prevent armed revolutionary Communists overthrowing the government. The first recruitment notice, signed by Noske of the Central Council, also appealed to the units to defend German borders, property and people against Poles and Czechs. The *Freikorps* units fought in the east in an attempt to maintain territorial gains in the Baltic; failed, then moved to the German-Polish border, where Polish nationalists were trying to expel German villagers from plebiscite areas. After the plebiscite, which was favourable to Germany, the Poles launched an attack on the disputed area. Some three-quarters of a million Germans were to be effectively expelled in the 1920s by Poland; their farms expropriated and 'compensation' paid in near-worthless government scrip.[17]

In early 1919, the militant Marxists, rejected by the vote of the Congress of Worker's and Soldier's Councils, staged an uprising, the Spartacist revolt. The Social Democrats used the *Freikorps* to put down the uprising. These units went on to break the power of the Councils. When in 1921 the Weimar Government began to put into effect Versailles clauses reducing the army to 100,000 men, two navy units under a Captain Erhardt led a putsch in Berlin. They declared a new government under Wolfgang Kapp, a civil servant, and General von Littwitz. After a few days the putsch faded away and Erhardt was imprisoned.[18]

The violence continued throughout Germany. Marches and counter-marches, demonstrations and counter-demonstrations, left hundreds dead on Right and Left in the first few years of the Republic. Each political party, the SPD, the KPD and so on, had its own storm-troopers.

In January 1923, France and Belgium invaded the Ruhr in an attempt to force the payment of reparations. The German currency, already weakened, collapsed. By October 1923, trillion mark notes were being issued. That year saw a peak of nationalist feeling and bitterness, while Golo Mann considers that anti-semitism was at its height just after the First World War.[19] Separatist movements grew in Bavaria and elsewhere. The re-unification of Germany was only thirty to fifty years old, and regional sentiments were strong. In the Ruhr the French backed a separatist group which instigated an armed insurrection. Hitler's November putsch was a by-product of all these circumstances. Propaganda harking back to the French occupation played a dominant role in the next twenty-five years.

Like many others, Darré found it hard to endure the period

immediately after the war. He attended a business school to learn book-keeping and shorthand, but felt ill at ease at home. His diary reveals resentment at the French street names and the presence of French soldiers. He missed the company of people of his own age and the comradeship of the army. Eventually, he decided to go back to the German Colonial School, and resume his agricultural training. There were many new applicants to the school, but a place was found for Darré because of his war record.[20]

By January 1919 he had met his future first wife, Alma Stadt, in Goslar, where she was at school with his younger sister, Ilse. He confided to his diary that as soon as he met her, he had an instinct that she would be his future wife.[21] Alma became the friend he longed for. He wrote to her several times a week, and expressed to her his sense of inner weakness. 'Look here—I will be absolutely honest with you,' he wrote in October 1919. 'My own idleness causes problems. Life has tossed me about so much already, has played so many dirty tricks on me. I've become mistrustful. Not only against fate, but of myself. I've experienced too often how I can become weaker and weaker ... I need a source of strength to lean on'.[22] Alma was not to be that source, and indeed Darré remained a solitary and mistrustful man, craving comradeship, but receiving instead either hero-worship or bewilderment from his peers. His letters to his wife over the next ten years oscillated between displays of affection and violent reproaches. He leaned on her, but resented what he saw as her stronger character. The tone of his communications shows a hectoring uncertainty: the need to teach and inform, but a fear of failure.

If one were to postulate a reaction against a stern father, who stressed the need for hard work, discipline and orderly business habits, one might expect to find a picture similar to the 1920s liberalism of England, after the Victorian sunset. Darré did in fact show many of these characteristics. His reaction led him to radical and revolutionary ideas, which, based on the humanism and republicanism of the Darwinian tradition, merged with the love of nature and the physical world inspired in late 19th-century Germany by Bölsche, author of *Love-Life in Nature*. He followed the cult of physical fitness. As early as 1905 the ten-year old schoolboy was winning prizes for swimming, and keeping a record of his achievements. He participated in running and swimming races at Giessen University in 1926, and sent the newspaper reports of his triumphs to his wife. He wrote to her about the virtues of the naked body. People resented nakedness, he argued, because the naked body expressed the truth of the inner person. You could not hide behind clothes.[23]

Darré tried to inspire Alma with his own enthusiasm for the cultivation and love of the body. He exhorted her to do gymnastics, and sent her a card by Fidus (Hugo Höppener), the *Jugendstil* illustrator, showing a naked man ecstatically greeting the sunrise on a mountain, his hair streaming in the wind. Their courtship remained platonic, though Darré wrote candidly to Alma, 'Physical love should be cultivated because it produces a certain release of energy (Entspannung der Energy) ... In nearly all marriages the key to unharmonious relationships is to be found in ignorance of these facts.'[24] The engagement between the two was kept secret for a year.[25]

However, as they were contemplating marriage plans, Darré suddenly had to leave Witzenhausen. A brief letter arrived for his father, on 27th November 1920, announcing that the student Court of Honour had found Darré guilty of lying. 'Some months ago I told a lie to protect the reputation of a comrade', wrote Darré. 'I have now been held guilty by the Court'.[26] The judgment meant automatic expulsion. There were no recriminations. The family flew to his defence, and a compaign began to have his name cleared of the offence. Although Darré's 1933 biography mentioned that he had taken his *Kolondiplomlandwirt* at Witzenhausen, he was not in fact able to take his final diploma at the time. He was also expelled from the regimental union of his old regiment, which hit him particularly hard. Vigorous support from the teachers at Witzenhausen ensured that the verdict was quashed, but it took nine years, and it was not until May 1930 that he was awarded his diploma. However, the affair seems to have aroused a competitive spirit in his academic work and attitude to life in general that had been lacking before. He went to Halle-Wittenburg in 1922, and despite distractions from politics, his marriage, also of 1922, and the hyper-inflation, ended by producing a good degree.[27]

Immediately after the war, he was still contemplating emigrating as a farmer to Argentina but was discouraged by the high price of land. The immediate aftermath of the First World War was a continuation of farm support prices and high land prices in Germany, fuelled by the incipient inflation, while South America enjoyed a boom from the world-wide shortage of food. Darré found that he would have to work for at least ten years as a labourer or gaucho before he could afford a farm of his own. 'It is a very healthy life', wrote his father doubtfully, 'but the work is extremely hard'. His heart problems, he pointed out, were hardly compatible with the brutally hard work necessary for a settler in South America. In January 1922, Darré contacted an Argentinian friend again, with a view to emigration, but received a discouraging reply. His last enquiry about Argentina was in 1923.[28]

But his next step after leaving Witzenhausen was to work as an unpaid farm assistant to gain practical experience. He began as a pupil on a peasant farm in Upper Bavaria; but the farm was sold after three months, and the new owner did not want him. Darré then worked as an assistant milker on a farm near Baden. Three months later there was a vacancy for an apprenticeship as farm manager in Gut Aumühle, Oldenburg, starting July 1921. Darré enjoyed working with animals—he had asked his father if he could run a stud-farm in 1917.

Darré found his experience of farm pupillage in Pomerania surprisingly rewarding. He decided to find a smallholding which would support his wife and himself, and his father-in-law offered to help finance him. He wrote to private and public land agencies looking for a small farm, but they were either unavailable, or offered at a prohibitive price. In some cases more than a million marks were being asked for part-shares in run down farms. In December 1921, Gut Aumühle changed hands, and the new owner brought in an old army friend, with no experience of farming, to take Darré's place. At this point, he began to make bitter references to the treatment of old soldiers in Germany, and the loss of German territory in the East. He noted that Admiral Horthy in Hungary had given his veterans land, and complained that the German plans to distribute farms to soldiers in the east in 1918, had come to nothing. In his second book, *Neuadel*, he was to stress the Horthy experiment.[29]

Darré's family had all been affected by the political and economic disasters. His aunt was now penniless, and decided to train as a teacher. His father seemed bewildered at the speed at which his prosperous, newly united country had lost, not only its colonies, but German territory. The German Colonial School had circularised its pupils' parents in late 1918, warning that economic recovery would be impossible without access to raw materials. The family funds were tied up in South America, though friends helped to smuggle some money to him. Darré's younger sister, Ilse, sent him a card in 1920 which showed a bearded peasant and his young son gazing across a wide river into storm clouds, while the peasant pointed out 'the Fatherland'.[30]

In 1922, Darré moved to Halle, to study for an agricultural degree at the Friedrich-Witte University, Halle. Here, he found a pleasant flat with a garden. He was constantly in debt, and Alma took over his financial affairs with his approval, and wrote to his father on money matters. He found his dependence on his father embarrassing, and wrote a long letter explaining that the war was responsible for his delayed professional training. His studies at Witzenhausen now appeared to be useless, since he could not use them as credits towards a degree in Halle. In Germany, unlike England, terms spent at one university can normally be set towards degree requirements at another. His younger brother, Erich was studying for a doctorate at Marburg. Like Darré, he was inspired by H. S. Chamberlain's work, and wrote, 'If I had time, I would like to busy myself with theories of inheritance and the racial question'.[31]

His half-brother was small and dark, a contrast to the four tall, fair-haired children. His successor's widow speculated that this may have helped to stimulate Darré's interest in racial questions. When his sister Carmen had her first child, Darré mused on the quantity of black hair the child was born with; it would soon go and be replaced by light-brown hair, which had happened with them all. Certainly, there was no black hair in the family, he pointed out. He seemed to have acquired an interest in medical matters from his father, who followed the latest scientific discoveries in biology, and recommended works on physiology and heredity when his son began his farm pupillage in 1921. When he began to study animal breeding, as part of his agricultural degree at Halle, his father again sent him book lists, and discussed 'Darwin and Haeckel's theory of evolution' with him.[32]

The collapse of his earlier plans left Darré in limbo. He spent much of his spare time reading a range of authors, from the great German classics to best-selling historical novels devoted to past German heroes who had struggled to conquer what was now lost. Darré read the classical *völkisch* authors, Houston Stewart Chamberlain, Langbehn and Lagarde. He also acquired a thorough grounding in German literature, reading Goethe, Schiller and Walter von der Vogelweide. Though never academically brilliant, he seems to have acquired a good general knowledge of German culture, specialised as his own training was. One letter home jokingly quoted a lament from von der Vogelweide when his money failed to arrive, another, criticising an attack on Protestantism by Count Keyserling, discussed Rousseau, Brentano, Eichendorf and others at some length; he was obviously well acquainted with the broad outlines of European ideas. Later he studied Sorokin, the sociologist, and read Nietzsche. He admired the ballads of Börries von Munchhausen, especially those dealing with the peasant soldiers of the eighteenth century. He read *Wanderjahr of an Engineer*, by Max Eyth, one of the nineteenth-century founders of the German

Agricultural Society, and novels about colonial adventures and farming. His reading of Lagarde's *Deutsche Schriften* and Lagarde's biography led him to criticise German education bitterly. He thought there had been a decline of general knowledge since the 1860s. 'How horribly empty today's education is.' He criticised the 'humanistic schools' for not teaching Latin, 'a sound foundation for other languages'. English should certainly be taught 'because of the USA ... though Yankee English is not exactly classical English ... Spanish is as important today'.[33]

His best work at Halle was in studies on animal breeding. This was carried out under a Professor Frölich, who was and remained a supporter of Darré and his ideas. He was later to suggest his name to the Nordic Ring, together with three other Professors who had taught him at Halle, as possible sympathisers. With the compulsory courses on agricultural subjects (which included meteorological observation, plant breeding, geology and chemistry), Halle offered courses on 'Racial Hygiene; its Possibility and Aims' (lectures given by Dr Drigalski, ex-army officer and member of the Society for Racial Hygiene), Social Policy, Economic Policy, the Versailles Peace Treaties, Plato, Ancient and Modern Philosophy, Hegel's Social Philosophy, Psychology and Medieval German Constitutional History. Mineralogy, plate tectonics and other physical sciences were also optional. Because of his studies at Witzenhausen, Darré was excused optional courses, but the range available is interesting. Just before his final exams, he sold an article, 'Animal Migrations; a clue to the Aryan Homeland', for 22.50Mk.[34]

After he had taken his farming degree at Halle, he spent a post-graduate term at the University of Giessen working on animal breeding with Professor Krämer, another academic with interests in eugenics and heredity. He was offered a post as administrator of a German farm settlement in South-West Africa; but he had decided that his future lay in Germany, and became more and more interested in nationalist politics. He now (1926) wrote his first influential political article, 'Internal Colonisation', which attacked 'dreams of empire', not only because Germany was unlikely to regain her lost colonies, but because he saw empire as inimical and destructive to the concept of the German homeland.[35]

He kept in touch with those of his former comrades and fellow pupils who did emigrate. This may have helped him to escape the desperate insularity and Eurocentricity of many other Nazi leaders. Later friends commented that he seemed more aware of the non-German world. One friend wrote from Mexico, 'Today I am a trader, and I have no idea what I will be doing tomorrow. In Germany people who change their professions are despised. In America, that attitude is unknown ... If you don't, you would be laughed at as stupid. In this respect at least American education is valuable.'[36]

It was rash of Darré to turn down the post in South-West Africa. By 1925 he had a farming diploma from Halle, but a doctorate was virtually essential to join even the Civil Service. His more highly qualified friends were now taking unpaid apprenticeships on farms while waiting for some public position to become available. With 50,000 agricultural civil servants already out of work, he found himself practically unemployable. The Wilhelmine bonanza for the civil service had come to an end. The Prussian Ministry of Agriculture refused to accept his practical farm work of 1921, and he thus failed to qualify for an appointment in the Ministry or even the County Agriculture Office. Now thirty-one years old, married and with a child, and without prospects, he had never lived with his wife for longer than a vacation, and the marriage already showed signs of strain.

When he was about to leave his first wife, in 1927, his younger sister reproached him with having become exactly like their father in his unfeeling egotism towards women.[37] The endless and nagging sense of duty imposed on the younger Darré by the elder produced a chronic sense of weakness and inadequacy in the younger man. The fact that his studies were prolonged unnaturally—long even by German standards—and that he was kept by his father and father-in-law for a decade, produced further defensiveness and resentment. His ideals were of manliness, honour and strength. In 1911 and in 1919 he had dreamed of being a soldier. Yet he found himself obliged to present his father with an account of his 'achievements' at the age of twenty-nine, and had no economic independence until the age of thirty-three. In 1919, when Darré was twenty-four years old, and after four years on the Western Front, his father wrote to reproach him for spending money on dancing lessons.[38] In later years Darré reacted by combining manic punctiliousness and military correctness in public life with personal charm and sensitivity in private life. Those who became his friends, especially many of his women friends, remained faithful for decades. In his political activity, his insecurity and self-doubt revealed itself in an almost obsessive cunning, plotting endlessly in the jungle around him, while he remained able to take unpopular long-term views on economics and politics with a visionary flair which was alien—and unwelcome—to those around him.

In December 1918, Darré's Argentinian pass, signed by the

Argentinian Vice-Consul at Wiesbaden, had shown a bony, oval face, with large blue-grey eyes, cropped fair hair, gaunt marks around the eyes and nose, and a full mouth pressed firmly together. By 1925, the sombre expression had disappeared, and his face had begun to take on the slightly fuller appearance it was to retain all his life.[39]

1925-1929

After his written exam at Halle in February 1925, Darré worked for a further term as 'Voluntary Assistant' under Professor Frölich until June 1925, when he took his oral examination. This was marked 'very good', and he thought of working on a doctorate at Giessen. During his semester at Giessen, in 1925-6, he again specialised in animal breeding, this time with Professor Krämer.[40]

Between 1925 and 1927, Darré published fourteen articles on animal and plant breeding, including works on the origins of the Finnish sheep, and seed breeding exchanges between East Prussia and the Baltic. This subject obviously pulled together many of his diverse interests, and, despite his late academic start, his work seemed to show promise. He hoped, indeed, to publish his Giessen work but failed.[41]

In June 1926 he was offered a post in East Prussia, as organiser and representative for the East Prussian Warm Blood Society—Trakhener Stud. The work was unpaid, but Darré would receive travelling expenses, and a visit to Finland was offered, to help organise a fair for the East Prussian Chamber of Agriculture. He was interested in the idea of working in the east. The practical and organisational work looked attractive, and he planned to work on a doctorate in his spare time. By now, he was thinking of supporting himself through journalism, as his first two political articles, 'Innere Kolonisation' and 'Rathenau' had been published, (although for a pittance: fifteen and twenty marks respectively) but the prospect of immediate involvement with a cause seemed attractive. The four years of war, his delayed studies, the disaster of Witzenhausen, meant that while he craved for and needed responsibility, he had no field of action. He had been a romantic, a lover of the outside life. Plans for colonial adventures had dribbled away through a mixture of ill-health, dilettantism and uncertainty. He was a comrade who could not trust his friends, suspicious and quarrelsome. He had striven to be a conscientious son to a father always suspecting him of dilettante attitudes, and failed, but now of his own accord began to look for involvement in practical affairs. From being disorganised and thoughtless, he became punctilious and methodical. From resenting control and discipline, he began to demand them from himself as well as those around him. Ambitions aroused, he believed his old faults were conquered, and his energies ready for use. If the East Prussian Studbook Society for Warm-Blooded Trakhener Horses wanted his services, he would throw himself into serving their interests, especially if he could combine that with serving Germany.

The story behind Darré's six months in the Baltic is hard to disentangle. The various biographical notices in the *Deutsche Führer Lexicon*, the *Neue Deutsche Biographie* and his 1933 biography all disagree with each other. His personal files show that he spent much of the time investigating an intrigue on the part of the commission house dealing with the Russians who were interested in buying the Trakhener horses with Prussian credits. Before discussing the possibilities behind his stay, it might be as well to explain the terms 'Warm Blood' and 'Trakhener'.

In the world of horse-breeding, there are cold-blooded horses, warm-blooded horses and thoroughbreds. Cold-blooded horses are for heavy work, warm-blooded horses can be used for riding. The famous warm-blooded breed of East Prussia was from the Trakhener stud, whose symbol was a double antler. It was a small, tough animal, with some English thoroughbred mixture. It was strong enough to work as a farm horse, but excellent as a cavalry animal. It was the fighting horse *par excellence*. In the 1920s, the Trakhener Stud sold thousands of brood mares and stallions abroad. Among the interested purchasers was the Soviet Government. The Weimar Republic, in 1922, had signed the Treaty of Rapallo, which agreed certain peace terms and trade agreements between Germany and Russia, while on 12th October 1925, the German–Soviet Trade Treaty was signed. These Treaties aroused violent opposition among the right-wing end of the nationalist spectrum (while the National Bolsheviks, together with figures like von Seeckt and General Schleicher, supported links with Russia as geo-politically inevitable). Under the Versailles Treaty, serious restrictions had been placed on Germany's diplomatic representation abroad. Because of this, Prussian ministries were known to use apparently innocuous agents and company representatives abroad as undercover negotiators and collectors of information—in short, as spies. The Prussian Ministry of Agriculture was noted for its tendency to lend itself to this practice.

Economic factors were involved, too. There was a surplus of grain products, and prices were low. Prussia was anxious to arrange exchanges of grain with other countries, and Russia, with her shortage of foreign currency, was the obvious choice for a barter agreement.[42]

The Trakhener stud sold some 5,000 horses to Russia during the 1920s, and the horses were paid for by Prussian government credits. Darré suspected the deal. He spent much of his time in Insterburg, where he was based from June 1926 on, in investigating what he thought was a corrupt situation, and sending nine-page long reports to his two immediate superiors. He received little response, but managed to annoy his employers by complaining at the lack of proper bloodlines for Hanoverian horses. He was later to compare the need to breed back to a pure Nordic strain to the need to breed Hanoverian horses back to a purer stock.[43] Darré also reported back on the bad economic situation for farmers in Finland, and the drought. It is hard now to establish whether or not the commission house was spying or cheating. Darré seems to have feared that Russia would pay for the horses by dumping agricultural goods in Prussia, to the detriment of the farmers. He was certainly present at negotiations with the Russians, and complained of dumping plans, and the incident reveals Darré's refusal to co-operate with established authority for the sake of his career. This attitude was not a helpful one in what was still a grace-and-favour society. When given an official ministry post two years later, he was besieged by requests to find posts for friends or their relatives. His situation was not helped by the fact that the Trakhener breed had always aroused unusual passions and loyalties: many of their owners were as combative as their horses.[44]

There is some evidence for the theory that Darré had an undercover role in the Baltic. In a letter to his wife in November 1926, he describes how he had gone to Finland to organise a fair for the East Prussian Chamber of Agriculture. Commercially, the fair had been a failure, although his organisation had been competent enough. The economic situation among Finnish and German farmers was too poor to persuade them to attend such an event. But, he added, he had had 'a special mission', which was 'a complete success', namely, to gather together Finnish and East Prussian farmers, with a view to forming an economic and cultural front, 'especially against Russia'.[45]

Was this on behalf of the Prussian Agricultural Ministry or had it been on his own initiative? Darré had been appalled at the poverty and misery among East Prussian farmers, especially the new settlers, most of whom had been agricultural labourers, and ran under-capitalised holdings, and was quite capable of trying to organise a nationalist resistance among East Prussian farmers on his own account. But in

February 1927 he told his wife that he was the 'German secret negotiator, to fight on economic matters between the Baltic and East Prussian Agriculture', which suggests that he was one of the undercover agents mentioned earlier.[46]

In 1927, Darré went to Lahsis, in Finland, as representative of the Agricultural Animal Breeding Association, and was asked to study the organisation of small dairy farmers in East Finland at the same time, after a report by a friend and supporter, Professor Carl Metzger, agricultural representative at the German Embassy in Helsinki, was sent to the Prussian Ministry of Agriculture.

In May 1927, he sent a despairing letter to Metzger, asking him to help find him a permanent post in the Baltic. He referred to his dependence on his father, and added. 'It may seem funny, but this is a very serious time for me, and I can't even afford clothes ... If I could go to Finland *officially*, then it would give me an inside position, and much greater scope ... This is the psychological moment for me.' A few days later, he heard that six new consular posts had been created, and in August 1927, he applied for a post as expert agricultural adviser and representative for the Baltic on behalf of the East Prussian Ministry of Agriculture. He was officially appointed in May, 1928, after months of knocking on doors in Berlin.[47]

Among others, he contacted Dr von Lösch, president of the German *Schutzbundes*, on the advice of Dr Edgar Jung, who told Darré that Lösch had 'great influence in the Foreign Office'. His plans for supporting East Prussian agriculture in the Baltic through barter agreements was laid before the President of the German Chamber of Agriculture, Brandes, by von Behr, who confided in Darré that Brandes was wholly in agreement with them. This was an important backer, as he headed the umbrella organisation of all German agricultural chambers. Darré acquired considerable knowledge of the highly organised structure of German agriculture during this period. For one year, indeed, he was a member of eight separate agricultural and eugenic organisations.[48]

In the summer of 1928 he was sent to Riga as representative of the East Prussian Agricultural Chamber. The intervening year was spent with his parents in Wiesbaden, where he wrote his first book *Das Bauerntum als Lebensquell der nordischen Rasse* (*The Peasantry as Life-Source of the Nordic Race*).

His years in the Baltic and the visits to Finland confirmed Darré's liking for the north German, and his view of them as a deeply rural people, threatened by internal decay and external threat. He referred in November 1926 to what he saw as his special role.

It does look to me ... as if my life's fate will be bound with East Prussia, as if God had called me to fight for East Germanness. I suspect that here in the East I might one day fulfil my life's mission, but we can only wait and see.

He was so absorbed in his work that in February 1929 he refused to return to Wiesbaden, in response to a telegram that his father was seriously ill. 'I can't leave the important negotiations with the Latvian Government,' he telegraphed back. Three days later his father died. Darré's comment was, 'When I think of the miseries of my father's marriage, my own reproaches are strangled.'[49]

His service in the Baltic ended with an argument with the ministry officials in Berlin over seed-breeding exchanges between Finland and East Prussia. He claimed the quality of the seed was poor. The matter was taken up in Berlin, where a court hearing found that Darré was right. However, he had aroused such animosity that he was recalled to Königsberg in June 1929 and waited some months for a new post to be found for him. He expected the national Ministry of Agriculture to take over his post, and offer him a new posting in January 1930. However, he had clearly lost favour. He claimed in his interrogation in 1947 that the political implications of *Das Bauerntum* led the Berlin ministry to refuse him service in East Prussia, and offer him a post in Ecuador instead but his 1933 biographer states that Darré resigned after Stresemann refused to accept proposals on agricultural trade between East Prussia and the Baltic States. This is possible, though there is no supporting evidence. In December 1929, he resigned his position. Later, wild rumours were to fly round the coffee-houses of Berlin that Darré had been sacked for illegal deals concerning Polish rye, that he was being blackmailed by a German industrialist who supported Darré as Hugenberg's successor. There seems to be no proof to support these rumours, which were retailed in the memoirs of a KPD man who worked for the NSDAP in the early 1930s; but their existence and propagation hints at the controversy and resentment that were to surround Darré in his political career.[50] He was later able to win support from the small, self-made businessman, who, like himself, was an outsider in Germany's complex interest group structure. Insofar as this was understood by organisation men of the right and of the left, it aroused resentment.

During these years, he began to show an ability to fire others with enthusiasm, to organise and to administer. He also continued to quarrel with most of his superiors, while managing to attract vigorous support from some. In general, Darré was capable of clear-headed, sometimes cynical, sometimes original, radical and persuasive insights. However, he was chronically suspicious of his confrères. This lends a sad and ironic flavour to his emphasis on the virtues of the English public-school spirit. At this stage in his life, these quarrels reflected not so much his combative nature as the mediocrity entrenched in German provincial administration.

POLITICS AND ANGER 1923-1926

In Spring 1925 the President of the Republic, Friedrich Ebert, died. Hindenburg, the 78-year-old head of the German armed forces during the First World War, was a candidate for the presidency, and was elected with a million-vote majority against a Communist and a centrist candidate, after the first election produced a stalemate. The election, like the revolution of 1918-19 and the insurrections, meant further violence. In Halle, the Stahlhelm took to the streets on behalf of Hindenburg. Darré took part in the marches and demonstrations. He described them enthusiastically, in an essay rejected by one journal.

> The Café Orchestra began playing the Erhardt song 'Hakenkreuz to the Stahlhelm', and there was a roar of applause ... The march was headed by nine young girls in gym costume. They were so clean and fresh ... It was inspiring to think that future German mothers would be girls like this with such healthy and supple bodies ... Why had 'Red Halle', blood-red Halle in 1919, become 'Red-white-blue[*sic*] Halle'? Because of the presence of over a thousand students of agriculture, energetic war veterans. The march surged forward unstoppably.[51]

In 1923, when the French and Belgian troops occupied the Ruhr, it was the Stahlhelm who sang the forbidden 'Deutschlandlied', and the old army song, 'We will hammer the French, and win' (Siegreich wollen wir Frankreich schlagen), a song whose refrain was, 'when you go away, we'll be there', phrases which recall chillingly the millions of corpses on the Western Front. 'In summer, 1922, it was risky to wear the Stahlhelm insignia in public, especially at the time of the Rathenau murder. Three years later, the Stahlhelm rules the streets in Halle.' Darré attended Stahlhelm rallies and veteran's reunions (banned, but held under the guise of a 'Prussian Day'), between 1922 and 1925.[52]

He was to comment in December 1925, when he moved to Giessen University for a term's graduate work, that it was a quiet unpolitical town, with none of the day-to-day political intensity and involvement of Halle, which was polarised between its large SPD and KPD working class, and supporters of nationalist parties. Darré had noted several times between 1922 and 1925 that 'red workers' were drifting into the 'fatherlandish' movements, but commented from Giessen that it was good not to be attacked in the streets by Communists just because you weren't a worker. 'While the Stahlhelm in Halle had lost five dead and badly injured during 1925, here you can ask the worker the way or the time of day without fear'.[53]

Darré had moved well to the right of his family by 1923, although within ten years his siblings and their spouses caught up, and became enthusiastic NSDAP members. He analysed the attempted *putsch* of November 1923 in Munich for his father, suggesting that it had been 'cleverly smashed' by the Bavarian Right to discredit Hitler. The left-wing press, he thought, blurred real differences of interest between right-oriented groups.

> If you stand on the left (by which I mean the left-wing of the People's Party—Centre Democrats—Social Democrats then Militarism-Ludendorff-Anti-Semitism-Von Kahr-Bavaria is seen as a unity, to be fought against as an ethical duty.

Darré attributed Kapp's opposition to the fact that Hitler's movement had support among workers 'contrary to expectations', that Hitler was trying to unite the *Sozial* movement, and had attracted support because Kahr was corrupt. 'Hitler may be a hothead, and not have a very political mind, but he represented a growing protest movement, that could not be crushed by force alone, in the way that Kapp's military *putsch* could be crushed.' Darré prophesied that it was essential to keep a close watch on 'this extraordinary current of ideas' in future months, years and decades, 'not because I want to defend it, or because it lies close to my heart, but because no one will understand Germany's internal politics if they don't.' He interpreted Hitler's working-class support as due to the economic failings of the republic, but 'I will be surprised if a man appears who can produce productive work for them, instead of destroying themselves and others in wild opposition.' Darré pointed out that the national army, under von Seeckt, had not supported the rebellion, and suggested that Captain Erhardt had been allowed to escape from prison in order to help crush Hitler in Munich.[54]

Was Darré more sympathetic to National Socialism than he was prepared to state in writing, especially to a disapproving father? In February 1923, worried about his lack of political involvement, he decided to express his feelings in a dramatic gesture, and, with a group of friends, joined the DVFP, or German People's Freedom Party. In March 1921, an agreement had been reached between Hitler and Von Gräfe, leader of the DVFP that the latter would have a free hand in north Germany. NSDAP members complained that the DVFP was insufficiently radical and too conservative. 'The poster inscribed "Jews and National Socialists are forbidden membership" was still fresh in our memory', wrote one Nazi farmer in 1937. Darré told his wife that the membership was growing rapidly and came mainly from the workers. 'The *völkisch* movement could never crystallise here as it could in Bavaria under Hitler.'

> So ... a new political party was founded on a purely national socialist basis, i.e. unceasing struggle against all Jewry and Parliamentarianism. ... In this I am now to the right of the German nationalists, and according to Wiesbaden ideas I belong in the madhouse, at least according to the *Frankfurter Zeitung*, which is the Bible with you,

he told his wife. His membership lapsed the next year, and appears to have been completely inactive. Darré failed to mention it in his NSDAP application in 1930, and always described himself as a latecomer to party politics. He failed also to mention it in his interrogation by the Allies in 1945, and while there were obvious reasons for playing down his early political involvement with the Allies, he could only have gained kudos by stressing it with his Nazi colleagues. His 1933 biographer claimed that Darré and a friend, Theo Habicht (see below, p. 75) had thought of joining the NSDAP in Wiesbaden in 1928, but had decided against it because of the damage it might do to their employment prospects. Certainly, Darré's wife joined the Wiesbaden branch of the NSDAP.[55]

Darré's letters to his wife between 1923 and 1925 often attacked what he variously called Jewry, the Jews and Jewishness. Yet in his published works up to 1929 there were no anti-semitic statements. Darré's position seems to have been that he was a political anti-semite, and felt no personal animus towards particular Jews. His position regarding Jews and race will be considered later in the book; this section will deal with his views in the early 1920s.

Many of Darré's attacks on Jews were because he associated them with democracy. There seem to be two separate points here; firstly, why did he so associate them, and secondly, why was he so anti-democracy?

Democracy in Germany at this time was not a buzz-word of approval; in fact, in some circles it was an insult, as, in earlier times in England, to call someone a Jacobite, a Tory or a demagogue was an insult. Indeed, Gladstone had protested when called a democrat, as he considered that the insult was so bad as to be outside the boundaries of acceptable parliamentary abuse. Many Germans outside the traditional conservative Right saw a democratic parliament as a means of abusing working-class power against the small businessmen, the farmers and the army; further, they saw it as essentially 'undemocratic', in that democracy was seen as a cover under which powerful vested economic interests could manipulate the population, a view entirely shared by the Communists. The popular press, the cinema, the legal system, were seen as a means of manipulation, not as a genuine expression of popular feelings.

Many modern studies of the 1920s have examined the extent to which the pre-1914 social structure was retained. Industrialists, generals and landowners are presented as engaged in a continuous and vicious fight against the frail, struggling forces of democratic virtue. There must have been, of course, industrialists who retained their possessions and fortunes through war, revolution and hyper-inflation; army officers who still held their posts, and landowners who had not been forced to sell their land for worthless currency in 1923. But the usefulness of physical assets as a hedge against inflation can be exaggerated. You cannot pay bills or buy bread out of bricks and mortar. The effects of the inflation were to render fixed incomes and pensions valueless, to bankrupt many creditors, and to interfere with internal trade. This, together with the effects of the war-time blockade, had meant years of hunger and sickness. In 1919, 90% of all hospital beds were occupied by TB cases. British observers in Germany such as Keynes commented on the starving children, the faces yellowed by shortages of fats.[56] One striking feature of photographs of German crowds in the 1920s—as in Britain in the 1930s—is the gaunt faces. Farmers were prepared to barter their goods, but reluctant to sell: what could industrial workers offer to men who would not accept worthless currency? Up to 1925, unemployment remained high, and the Weimar government's attempt to palliate conditions for the urban workers by means of high welfare benefits had to be paid for, in part, by taxing the farming community. This inequality of tax burdens was a major factor in farming support for the NSDAP in the early 1930s.

German nationalists held democracy responsible, and tended to consider 'Jewish democrat' to be one word. Certainly, there were Jews among the new government, and some commentators, like Thomas Mann, saw this as an index of a new humanity and tendency to socialism.[57] The statistical disproportion was hardly surprising. Conservative and monarchical structures tend towards exclusivity; in Europe such structures traditionally excluded Jews: not always literally, as legal constraints against Jews in the professions, in land-owning and in the non-Prussian army were abolished in nineteenth-century Germany. But many successful Jews felt that they were not fully accepted, although a kind of parallel hierarchy developed in the social and economic expansion of Wilhelmine Germany. The social democratic governments that emerged from the collapse of the European empires in the early 1920s contained a high proportion of Jewish members, as did the Marxist revolutionaries of Russia and Hungary, not to mention the leaders of the revolutionary councils in Germany. Russian Jews were the most likely of the Russian minorities to have the language skills and foreign contacts necessary to become ambassadors to and negotiators with other countries, and were thus more noticeable. There are, of course, methodological problems in making generalisations of this kind, but they are problems unique to this area. If Croats had emerged in prominent positions in several European nations during the 1920s, it would have been a matter for considerable comment. If the movements associated with prominent Croats were seen as hostile to traditional nationalist values, it would have been commented on adversely. German nationalists, especially the less conservative ones, reacted in this way to the prominence of Jews in democratic and socialist movements, while nationalist German Jews seemed to switch their allegiance from the lost Wilhelmine empire to Zionism in the 1920s.

Since the Second World War, people have become infinitely more sensitised to the issue of labelling people or groups in a pejorative way, and especially when those concerned are Jewish. Generalisations have continued to be made about other groups, tribes, nations and races, both in historical and scientific works and in everyday life. But there is a convention that it is perceived as anti-semitic to label someone a Jew if the context is pejorative, but methodologically sound to label someone a Jew if the context is favourable. For example, writers in the 1920s (including Keynes) who labelled the early Soviet leaders as Jewish have been attacked as anti-semitic; but Trotsky could be labelled as Jewish once he began to oppose Stalin. Einstein is usually labelled as Jewish, although not a practising Jew; but anyone writing about 'the Jewish criminal Stavisky' would be perceived as anti-semitic. This leads to problems in understanding the mentality of 1920s Germans. If one were to write that Jews have always been in the forefront of movements designed to liberate the poor and oppressed, that they have always taken up the cudgels on behalf of trade unionists, idealistic internationalists and victims of superstition, one would be saying, objectively, exactly what the *völkisch* writers of the 1920s were saying; that Jews supported Socialism and atheism, and opposed conservative nationalism. For the historian, a more important aspect is, not whether these accusations were anti-semitic, but whether they were true. Only then can the perceptions and motivations of the time be understood.

Darré, in his anti-semitic pronouncements, was not only supporting conspiracy theories, but trying to construct a model of the future, insofar as it would affect him and his family. He thought that Jews in various countries would automatically act in support of their fellow Jews elsewhere, and would automatically change their national allegiances when they felt their interests threatened. He thought that Jewish interests had brought America into the war, because they saw that 'Albion's star was sinking', and that only America and Germany could stand alone: they chose to support America. He asserted that the Jews supported a United States of Europe because 'seen from the financial-technical—i.e. Jewish—point of view' this was the most efficient way to control industrial production; the Dawes plan was a means to that end. Darré thought that the *völkisch* movement was the one danger to this plan, but that the 'farcical failure of the Hitler *putsch*' meant that it was impotent. At the time of the Locarno Pact negotiations, he argued that France was waiting for Germany to collapse. He admitted what he saw as Jewish sense of identity and tribal feeling, but resented it too. 'The Jews think they have the right to build up land in Palestine as much as they want, but we in Germany must not do so'. He argued that Jews wanted to deny nationhood to others because they saw it as a danger to themselves, but claimed the right to nationhood. He ended a tirade against 'this damnable concept of internationalism ... Marx, Lasalle, both Jews' with the prophetic comment:

> The choice is now between nationalism and internationalism; that is, between anti-semitism and pro-semitism ... and I think I can safely say that 90% of students are anti-semitic.

Modern scholars have agreed with this estimate. During the 1920s, Jews were excluded from many student associations. One such association was refused a grant from the Prussian Ministry of Education unless it accepted Jews, but preferred to forego the grant rather than change their rule. The fact that ministries opposed anti-semitism among the young enhanced its attraction.[58]

In short, when Darré associated the 'Soviet star' with the Jews, he was expressing a widely held view of the time. He thought the post-war economic chaos a result of trying to enforce the international debt system, and associated manipulative democracy with Jews. His main attacks in his letters were reserved for France, but it was so obvious to him that France was an enemy that he did not blame her for her hostility. He decided that the nationalist movement in Germany was the only movement prepared to defend Germany against France, and organised Jewry, inside and outside Germany, was the undying enemy of this nationalism for power-political reasons. This was a political anti-semitism which should not be confused with his Nordic racialism.

The 1920s certainly saw an attempt to build a new economic international order, with sporadic attempts to incorporate two 'rogue' powers, Germany and Russia. From January 1924, indeed, Germany knew stabilisation and economic growth to a remarkable extent. The new mark was introduced, and an American banker, Charles Dawes, evolved a new system for reparations. This was known as the Dawes Plan, and recommended that reparation payments be balanced by Germany's ability to pay each year; and was followed by American capital flows into Germany. Unfortunately, this early example of debt rescheduling ran into problems. The capital was short-term money, funnelled into long-term investment. When the American crash of 1929 came, the money was withdrawn, leading to the near-collapse of the German economy. By 1932, six million men were unemployed. The Dawes plan had been attacked from the first by German nationalists, firstly because it accepted the idea of reparations, and secondly because it was seen as a means of extending the hegemony of New York bankers (and, by extension, the Jews). Every attempt to negotiate with other nations by the Weimar Government was attacked as a betrayal of German interests.

The rapid political changes, the loss of territory and 'face', the economic misery and uncertainty, went along with an explosion of intellectual discontent. 1920s Germany was in many ways similar to a Third-World post-colonial society in its mentality, existing institutions seen as alien and hostile. Anything was possible given the human will.

Nothing was possible, given human inadequacies. In this ferment, Darré began to develop his extraordinary package of ideas. Like a more nationalist Che Guevara, he opposed capitalism and the town. He sought a new form of corporatism, in which farmers would govern themselves, and factories become unnecessary. Most elements of his views were not original, and some of their sources will be discussed in the next chapters. What was original was the fusion of these elements into a unique combination of messianic despair and voluntarist optimism. This gave him his force, his following, his brief triumph and eventual inevitable failure. The force of the ideas developed in the 1920s drove him to seek office in the 1930s. It was to prevent him from compromising with Hitler, and led to his downfall.

CHAPTER TWO

Odin's Ravens

We live today in an age when the desire to return to one's ethnic and cultural roots is taken for granted. Even the most internationalist intellectuals make exceptions in their dogma for the fashionable minorities; the Bretons, the Basques, the Orcadians, the blacks, and threatened, or supposedly threatened Third World tribes. There is a tendency to assume that majority groups cannot be threatened, and therefore cannot attract the sympathy or interest of the observer. Majorities are out; minorities are in. Volumes have been written about this strange intellectual quirk, and this is not the place to try to investigate it. It may, however, be useful here to remind the reader that most groups can, objectively speaking, be threatened, whether majority or minority. Germany for much of the 19th century was a potential majority scattered over and among minorities. At this period the 'nationalities question' meant the right to national self-determination of Germans, Magyars, Italians and Poles, four major European groups whose ethnic boundaries did not coincide with their political boundaries. The search for roots was not confined, in short, to the German romantics. The discovery of ancient graves complete with bodies helped to inspire an interest in physical anthropology, and in France and Scandinavia, history was now seen as the result of conflicting racial and tribal forces. Who were the people of Bronze Age Europe? Where did they come from? How were language groups related to ethnic groups? What was the pattern of settlement and colonisation of Europe?

By the late nineteenth century, new self-conscious groups were emerging. These included the Balkan states, and those Slav nations invented by the agile minds of a handful of Oxford historians. They also included the Scandinavian countries of northern Europe, the Nordic nations. Nordic has become a word associated so exclusively with Nazi propaganda that it is always a surprise to encounter it as a value-free description of Denmark, Iceland, Norway, Sweden and Finland, which is how the word is still used. Originally, the word simply meant northern, and became commonly used to describe the tribes of northern Europe, along with *ostische, westische* and *südische*, although it was temporarily substituted for 'Aryan' in the 1920s. Nordic was not co-terminous with Aryan, Indo-European or Germanic. Aryan originally referred to the tribes of the Iranian plains, and was used as such by Darré in the 1920s. 'Indo-European' referred to a language group.[1]

Some definitions of 'German' seem absurdly wide, as when H. S. Chamberlain used the word to mean all non-Mediterranean and non-Tartar peoples (i.e., to include, 'Germanic Celts,' and 'Germanic Slavs'). This was the racial definition, used as a catch-all to ensure that famous German heroes, such as Luther, did not slip through the net. The later categorisation of northern Europeans into five or six sub-races cut across national boundaries, but this time cut across existing concepts of Germanness, too. As was pointed out at the time by anti-Nazi journals, many Nazi leaders were themselves non 'Nordic'. However, the common notion that all Nordics were supposed to be blonde is misconceived.[2]

Northernness had its appeal in England, too. English nationalists and populists were inspired by the world of the sagas and Norse myths. William Morris, J. R. Tolkien and C. S. Lewis are among those who looked for inspiration to the treeless hills of the north, the wide, sandy dunes and the chilly seas, in part because of the appropriation of the natural and the spontaneous carried out by Arnoldian celtophiles. In his autobiography, C. S. Lewis described his yearning for 'Northernness', and his reaction on opening the Rackham illustrated *Twilight of the Gods,*

> ... this was no Celtic or sylvan or terrestrial twilight ... Pure Northernness engulfed me: a vision of high clear spaces hanging above the Atlantic in the endless twilight of Northern summer, remoteness, severity ... the same world as Baldur and the sunward-sailing cranes ... something cold, spacious, remote.[3]

In 1922 a work called *Rassenkunde des deutschen Volkes* (Racial Handbook of the German People) was published by Hans Günther, and became a best-seller. It differed from earlier popular works on race and culture in its deliberately scientific attitude and analytical vein. The *völkisch* writers of the late 19th century had concerned themselves more with art, culture, civilisation and spirit than with quantitative physical distinctions between groups. They wrote in the tradition of German idealism. Günther, working from recent developments in physical anthropology and eugenics, argued that mankind was divided into races which differed in physical structure and mental character. The highest type was the Nordic, comprising the populations of Scandinavia, north Germany, Holland, Britain and the United States. The inclusion of the United States may seem surprising, but the USA was still seen by some as an Anglo-Saxon country. For example, Lenz, co-author of a text-book on biology, had called on the USA to lead a 'blonde international' after the First World War.[4]

The Nordic movement that arose in the 1920s was first a pan-national and a cultural one; it searched for ancient northern myths and sagas, and, especially in Denmark and Germany, it emphasised the rural nature of the Nordics—in fact the peasant adult education movement known as the *Bauernhochschulbewegung* had its origin in Denmark. The Nordics were seen as racially and culturally threatened. They were surrounded by other groups, split between nations, without their own territory or unity, and their birth-rate was declining, as was the German growth rate since 1912. However, most Nordicists jibbed at supporting breeding communes of the kind suggested by Willibald Hentschel in *Varuna* in 1907. The close, monogamous family was seen as essentially Nordic.[5]

Despite his scientific attitude to physical, measurable differences, Günther's analysis of *mental* characteristics was superficial, in the anecdotal, descriptive fashion rendered respectable by Max Weber.[6] Nordics were reserved but trusting, creative, had a sense of justice, and were good fighters. Mediterranean peoples were lively, unserious, charming, etc. A description of physical differences between groups and races is still part of physical anthropology, but scientists on the whole have been relieved to leave to journalists generalisations about the spiritual qualities of different groups, except where they relate to quantifiable elements, such as differences in glandular secretions or brain structure. This is not to say that sensitive, perceptive comments about groups have no role to play, still less that they should be banned, but it was unfortunate for Günther and other racialist writers that they tended to confuse their own common-sense perceptions (notoriously a bad guide for scientific comment) with the results of carefully conducted surveys. In the case of Chamberlain, who in his later years abjured Darwinism in favour of a vitalist philosophy, his anti-semitism seemed to lose its strictly racial content, and become more culture-oriented.[7]

For the German Nordicists, Scandinavia was seen as the rural ideal. Hans Günther married a Swede, and lived in Sweden during the 1920s for some years. Scandinavian writers were popular. Knut Hamsun and

Selma Lagerlöf depicted a stern, hard-working, but deeply satisfying peasant society. The peasant was strong, inarticulate, but superior. In Scandinavia, where there were very few Jews, there was little anti-semitism in the Nordic movement; instead, animosity was directed to the petty bourgeois townsman and bureaucrat as the enemy of the peasant. The town meant administration, taxes, small-mindedness and tuberculosis. Adventurous industry was acceptable: it meant opening up the land with roads and railways, and scope for individualism.[8] Anti-semitism was scarcely visible in the published writings of the German Nordicists, but certainly the Jew was identified as 'the enemy' of the Nordic movement in Germany, something that emerges more clearly from their private correspondence.[9]

The exclusive and tribal character of the Nordic movement meant that Hitler's National Socialism could never fully absorb it. The Nordics offered Northernness as the *best*, the most important part of the German heritage. In doing so, they excluded the Catholic parts of Germany, including the blonde, blue-eyed German Catholics of Austria and much of south Germany, also the Rhineland states. The Nordics preferred Protestantism to Catholicism, because Protestantism was seen as the Northern reaction to an alien Christianity, a move back to the purer, more individualistic spirit of the north. The National Socialists, on the other hand, especially in office, offered a movement of German unity above particularist loyalty. When Darré started a group in 1939 to study northern customs and report on eugenic questions, he did so as a deliberate gesture of defiance, and in conditions of semi-secrecy. One of the most important early Germanic racialists, Lanz von Liebenfels, had his writings banned in 1938 while other occultist racialists were banned as early as 1934. Schultze-Naumburg's 1930 plans for a 'high-quality racial core' were also laid in secret. The fissionable nature of Nordic thinking was recognised and feared by Goebbels, certainly, well before 1933.[10]

The main eugenic interest among Nordicists was to preserve what they feared was a dying group. It was a defensive concept, a fact which is obscured with talk of 'Supermen' and 'Master Race', phrases which are rare in Nordicist writing. What was at issue was not the breeding of a *superior* race, but the preservation of a threatened race. This theme of Nordic vulnerability was to be prominent in Darré's work, especially where the Nordic peasant was concerned. Here, the threat lay not only in a decline of the birth rate, economic hardship, cultural decay, but in urbanisation.

Darré began to read Günther in 1923, and in 1926 was given an immensely popular *völkisch* work, Julius Langbehn's *Rembrandt als Erzieher*. Langbehn was interested in cultural history more than scientific racial questions (the first, 1890 edition of his book contained no mention of the Jews, although in later editions, two anti-semitic chapters were added). He combed European history for great Germans, or great men of German stock. 'German' was always a fluctuating and difficult historical and ethnic concept, and it was easy to write Germanness out of European culture by re-labelling. The borders of the German states and confederations had fluctuated so much that German identity was in a state of permanent confusion. There was a sense, especially among intellectuals, that their moral and psychic space was insecure. There was a national inferiority complex. Not only did political parties after 1871 include the word German in their formal titles; the full title was always used; the *German* National Party, and so on, indicating that national identity could not be taken for granted.[11]

Ernst von Salomon, the conservative Weimar intellectual, writer, and former *Freikorps* member, wrote in 1951 that, born in Kiel before the First World War, he was a Prussian, not because he was born in Schleswig-Holstein, then under Prussian rule, but because of his ancestry—although his father was born in St Petersburg, and his mother in England. He explained that this was because he had no 'biological connection' with Kiel, St Petersburg and England. This comment encapsulates one of the main distinctions between German identity and that of other European nations. In England, for example, nationality had a territorial quality, but historically, the German concept of Germanness was so little linked with territory, that the 1848 liberal Frankfurt Parliament had proposed to give the vote to all German émigrés living outside Germany, be it in America, Russia or elsewhere in Europe. The identity problem was exacerbated by the fact that the two largest and most powerful German states, Prussia and Austria-Hungary, both had substantial non-German populations; in fact, they were multi-racial entities. From Herder onwards, 'Germanness' had been defined largely in terms of language. What Langbehn was trying to do was to define Germanness in terms of ethnicity.[12]

He set out to establish an idea of Germanness that was to be found in art and culture over the centuries, wherever people of German ethnic stock had settled. He located the full flowering of this Germanness in Rembrandt, who came from the Netherlands, a point not lost on Langbehn's nationalist critics. Langbehn was uninterested in party politics and the German Reich, and was attacked for implying that the Prussia–dominated German state after 1871 was in some way inferior to the Netherlands. 'Will Germany rule the world if every German becomes a Rembrandt? Will this rescue us?' asked an anonymous author in the 1890s. 'Holland and Switzerland are presented as ideal states, but what would become of German power if this idea were carried through?' This comment encapsulates the opposition between the 'Greater Germany' answer to a search for German identity, and the power-political, common-sense answer, which was to rely on a Prussian-dominated Germany. Langbehn's weakness lay not so much in the idea behind his work, unpopular as this was with many powerful groups in the thrusting expansionism of Wilhelmine Germany, as in the wild over-simplification of his pronouncements. He idealised the peasant, seeing him as the true aristocracy. This idea at once coincided with and codified Darré's thinking.

> The Lower German (*Niederdeutsche*) is a born aristocrat in character. No one is more recognisably aristocratic than the true peasant, and the Lower German is always a peasant. The peasant, like the farmer of North America, the English Lord, the old Mark nobility and the South African Boer, belong spiritually and physically to one and the same family,

he quoted to his wife. Langbehn cited Cromwell as the typical peasant (*sic*) turned king, whose eyes were 'grey and fearsome as the North Sea'. Darré criticised Langbehn for being one-sided, and found Chamberlain and Günther more inspiring, possibly because they were less concerned with cultural and spiritual elements.[13]

Along with his reading, and a sense of rootlessness, went a growing obsession with his own ancestry. His sister Carmen shared his interest, and visited Sweden to track down their ancestors there. He was pleased to find a noble Norwegian ancestor in the thirteenth century, as well as sixteenth-century French Darrés, and addressed the Halle Genealogical Society several times while a student.

He described Günther's *Rassenkunde* as 'wonderful'. He meditated, 'I feel tremendously drawn to the Lower Germans and Lower Saxons—am probably more Lower Saxon than really Nordic in Günther's absolute meaning of the word.' He told his wife to soak herself in Günther's book *Style and Race*, so that their children could be brought up 'Nordically'.

> Good, 100% Nordic children cannot come from our marriage. But you could bring them up in such a way that their Nordic

blood, in so far as it is prominent, feels at home in their upbringing.[14]

Darré found in Günther's highly complex and finely tuned categorisations an explanation for his unhappy home as a boy. His parents had failed to cope with the problems caused by his 'mixed blood.' He commented in his second book, *Neuadel*, on his sense of uprootedness and search for Germanness. 'One kills the German soul if you remove the countryside where it was born ... For a child brought up in the terribly cold spirit of the American milieu, it is impossible to understand German myths' tales and legends.' His younger brother Erich, now working on a doctorate at Marburg University, he described as 'a Nordic dreamer through and through', who also had not learned to cope with the problems of the Nordic heritage. 'If he had been brought up in Nordic circles he would not have got mixed up with coffee-house intellectualism.' He concluded that many of what he had always considered personal weaknesses of his own character and manner (dreaminess, indecision, anger) were probably the result of his heredity 'although one can escape from some of these characteristics', he added, with that puzzling mixture of a belief in heredity and in will-power and environment, that was to characterise his later political views. For example, at this time he described Prussia as a *successful* mixture of types and races, held together by political tradition and will, a position he was later to oppose.[15]

His article on 'Internal Colonisation', which will be discussed in the next chapter, was followed by a *jeu d'esprit* on Walter Rathenau's pre-First World War racial concepts. Rathenau was the millionaire Jewish son of an industrialist, who became Foreign Secretary under Ebert and was murdered in 1922 by a nationalist group. One of their grievances, according to Salomon, an accessory to the murder group, was that Rathenau, as a Jew, had no right to be a German nationalist, and further, was a fascist. Rathenau had been influenced by Germanic racialist ideas, which he combined with the peculiarly liberal imperialism of the type associated in Britain with Joseph Chamberlain and Cecil Rhodes.[16] The article began by announcing that Nordicists were wrong to attack Rathenau, and gave extracts from Rathenau's pre-war work *Reflections*, such as, 'Voluntary, instinctive respect rests completely on racial perceptions. One would rather obey a noble, white hand than clever arguments.' Or,

> The task of the coming time will be to breed and to train up the dying or impoverished noble races needed by the world ...

Man must follow this course deliberately, as used to be done by Nature, the course of 'Nordicisation'. A way of life which is physical, full of endurance, severe climate and solitude.

Rathenau stressed the need to breed a new ruling nobility from Aryan stock, and talked of exalting pure Nordic blood, and imitating natural eugenic processes in order to produce a stronger Aryan race. Race, eugenics and new nobilities were in the air both before and after the First World War. For example, in 1921, the son of a Prussian nobleman and Japanese mother, Count Coudenhove-Kalergi, wrote a pamphlet, *New Nobility*, in which he advocated the breeding of a new master-race which would rule Europe. *Two* master-races already existed in Europe, the Germans and the Jews; they should join to breed an unbeatable super-nobility, he argued. Despite this somewhat unexpected conclusion to his argument, Kalergi was on the invitation list for the National Peasant Council meetings during the Nazi regime, although it is not known whether he attended them (he was then in Switzerland). Kalergi, whose grandmother was a friend of Houston Stewart Chamberlain, was a founder member of the European Movement after the Second World War.[17] Another supporter of a joint German–Jewish aristocracy, fighting democracy together, but not interbreeding, was Dr Oscar Levy, editor of Gobineau and Nietzsche, who reproved Gobineau for not paying enough attention to the pure blood of the Jewish nobility. Levy was to defend Nietzsche after the First World War against the charge that he had inspired German soldiers.[18]

Darré knew perfectly well that Rathenau was merely expressing the intellectual fashions among ambitious Wilhelmine businessmen, but, tongue in cheek, he announced his desire to show that Jews could be splendidly racialist. The article caused a stir among Nordicists, and was immediately reprinted, without permission, by smaller periodicals. He followed it up two years later by a further collection of racialist quotes, adding this time Disraeli's well-known pronouncement from *Coningsby* that 'Race is the key to world history'.[19]

The Rathenau article was important because it introduced Darré to Nordicist circles, and people who were to play a major part in his life. They included the founders of the Nordic Ring, a loose collection of Nordicist groups, which aimed at coordinating all the Nordic movements, established in May 1926 by Hanno Konopacki-Konopath, a top civil servant (Ministerialrat), and his wife, Princess Marie Adelheid Reuss zur Lippe. They published *Nordische Blätter* and Konopacki was later to edit *Die Sonne*. Darré and Marie Adelheid promptly formed a close political alliance, and became personal friends. Darré called her his 'Little Sister', and was godfather to one of her children. She edited several collections of Darré's writings. He lectured her on eugenics, and sent her reading lists.[20]

Other members of the Nordic Ring included Erdprinz zur Lippe, later to work for Darré, and a member of the 1939 Society of Friends of the German Peasantry, and a Baltic German, Freiherr von Vietinghoff-Scheel, an émigré to Germany who lost his lands in the Baltic to the Russians, and later lost his farm in Germany through the Depression. His daughter, Charlotte, became Darré's secretary in 1929, when they stayed with Paul Schultze-Naumburg in his home, Burg Saaleck, and she was to be his second wife. Vietinghoff-Scheel was the author of *Grundzüge des völkischen Staatsgedankens* (*Foundation of Populist National Constitutional Ideals*), which demanded the exclusion of Jews from public life.[21]

Charlotte Vietinghoff-Scheel is remembered as a polished, elegant, well-educated 'gentlewoman' (it is interesting that the English term was used). Frau Schultze-Naumburg was a match-maker. The young daughter of a Baltic nobleman was a better companion for their admired friend than the unseen, difficult wife, living in semi-separation. Darré, who missed the close contact with his own brothers and sisters, now married, liked Charlotte's family. His jokey, sketch-scattered letters now went to Karen von Billabeck, Charlotte's sister, as well as to Marie Adelheid. He found Charlotte, who was ten years younger than he was, attractive and companiable, and fell in love.

They became secretly engaged, and he wrote to tell Alma. The marriage had more or less broken up in 1927, partly because of the strain of constant financial problems, the separations, and partly because of Alma's jealousy over Darré's friendships with other women. Although he did not have the reputation of a womaniser, he found it easier to form close and affectionate relationships with intelligent, lively women than with men. The daughter of the landlord at Gut Aumühle had continued to exchange letters with Darré for the next ten years, for example. Also, Alma, while a strong-minded woman, and thus the 'source of strength' that Darré wanted, had known him through the days of his failure; now that he had found a niche, he seemed to develop the confidence to break away.

He left the decision about a divorce up to Charlotte (known as Charly), but told Alma that he regarded Charly as his real wife. Nonetheless, he mused wistfully on the Japanese habit of having first *and* second wives. Alma tried to understand exactly what she had failed to supply. Had he been happier at Saaleck, she asked? Darré responded that it was not happiness that he had found, but an inner freedom: he now felt able to draw on his own strength. It is unclear whether politics had anything to do with the break-up. Charly's father, who belonged to the Nordic Ring, was probably more sympathetic to Darré's views than the hard-headed merchant, Jacob Stadt, Alma's father. But Charly, despite the fact that she had been unconventional enough to become secretly engaged to a married man and, once divorced, married him—surely unusual behaviour for a well-brought up girl of good family in those days—remained unconnected with politics, apparently through reticence. She seldom appeared at public functions, and did not involve herself in party matters. Yet she obviously provided some strength and inspiration for Darré to draw on, and the marriage was a close and happy one. He still continued to write to Alma until 1931, and the correspondence was resumed in 1945, until his death in 1953.

Darré was first contacted by Marie zur Lippe, asking him if he supported the Nordic Movement, and suggesting that they publish his Rathenau article. He replied doubtfully that his interests were in 'scientific animal breeding', and that he did not really have anything to offer the Nordicists. He did send her a list of those of his former teachers at Halle and Giessen who were potential sympathisers, and interested in 'racial science'; Professors Krämer, Frölich, Römer and Holdefleiss. Konopath, however, refused to contact them on Darré's recommendation, until he knew more about their stance. 'Mere racial hygienists are not always on our side', he noted. 'Professor Pohl of Hamburg, for example, although a racial hygienist, is half-Jewish'.[22] While there was a tendency to ascribe Jewish blood to anyone whose ideas differed in any detail from the party-line Nordicist, Konopath's attitude is a useful reminder that racial hygienists, eugenicists and anti-semitites were not co-terminous.[23]

Marie Adelheid had signed herself 'with Nordic greetings'; but Darré ended his reply with 'German greetings', the first time he used this nationalist form, and a gesture that signified the secondary role Nordicism was to play to peasant problems for him. By late 1928, in *Neuadel* (p. 29), he reverted to the term 'Germanic race', adding, 'or the "Nordic" race, according to whichever expression is fashionable.'

The Nordic Movement was based in Berlin and other parts of North Germany. One Nordic Conference held in June 1930 (or Brachmond, according to the *völkisch* calendar then in vogue), was fairly typical. Hans Günther, by that time appointed Professor of Racial

Hygiene at Jena by Frick, the National Socialist Education Minister of Thuringia, spoke on the education of youth in Nordic thinking. Darré summarised the ideas of his book in a talk entitled 'Blood and Soil as the Nordic Race's Foundation of Life', and Schultze-Naumburg gave an illustrated talk on the Weimar University of Architecture, Arts and Crafts. The Conference was attended by members of the Nordic Ring, the Fighting League for German Culture, the German League, the League for German World-View, two university groups, the Northern Order, the Nordic Youth Group, the Society for Northern Beliefs, and the German Richard Wagner Society. The emphasis was on youth and culture. One group that attended, the *Ahnenerbe* League, was later to be incorporated into the SS. None of these talks concerned eugenics or racial hygiene. However, a strong racial feeling underlay the cultural and educational interests.

> The Nordic Ring considers its first and most important task after the bringing together of all Nordically minded people, and the creation of a racial nucleus (*Rassenkern*), the planned propagation of Nordic ideals.

wrote Konopath on the foundation of the Nordic Ring.

The group held meetings once a month. Between 1929 and 1930, they heard talks on Viking battles, Darré on old Nordic family customs, the Nordic world-view, and the Berlin Theatre. Marie zur Lippe led a discussion entitled 'Is the Nordic Idea materialistic?'. A speaker from Oslo gave an illustrated talk on 'The Effects of Civilisation and Racial-Mixing on National Strength'.[24]

Most of the speakers were German, with a scattering of members from Scandinavia. Only one speaker came from Britain, a German linguist who had been a member of the Allied Commission after the First World War, and apparently stayed on in Germany. He wrote an attack on the French behaviour in Germany in 1929, and in 1933 was to greet Hitler's rise to power with the call 'England erwache'.[25]

There were Nordicist groups elsewhere—such as the well known circle around Johann von Leers, who then lived in Munich. Darré met him in 1927, and kept in touch. Dr von Leers spoke fluent Japanese, and later became a professor at Jena University. He seems to have worked for the German Labour Front up to 1939, and to have been an SS member. Darré wrote to Hitler to vouch for Leers' reliability in 1939. His wife believed herself to be the reincarnation of a Bronze Age priestess, and held regular meetings at their house, where she would wear barbaric gold jewellery. Other guests included an Austrian who had renamed himself *Weissthor*, the White Thor, to the irritation of some of the more aristocratic guests, who pointed out to each other that Weissthor's real name was more plebeian. George Mosse, in his book *The Crisis of German Ideology*, rather tetchily criticised von Leers for still being a sun-worshipper in the 1950s, as if the experience of Naziism should have put anybody off sun-worship. But the rejection of Christianity and search for Nordic myths and religions long pre-dated National Socialism, and outlasted it, while Theosophy and Anthroposophy should also be seen as part of this rejection of organised religiosity. One consequence of the quarrying of pre-Roman northern European customs was the formation of Druidical groups in Germany. Ironically, the German Order of Druids was caught in the National Socialist anti-Masonic law of 1935, and was closed down, despite an appeal to Darré, protesting to the last that they were not Freemasons but good, German Druids.[26]

Darré brought a new element into the Nordic movement. His training in agriculture, his farming experience, his animal breeding, his liking for evidence and argument, and his capacity to inspire enthusiasm and activity all galvanised the movement. He may have been seen by the later Nazis as an unworldly dreamer, but compared with the Nordic Ring he was tough and practical. At least, this was the role he adopted; the plain man of science, the sensible, forthright polemicist. He blossomed among his new friends. He moved among them as an outsider, but, perhaps for the first time in his life, as a superior outsider; more experienced, practical and worldly. He began to write lively and charming letters, more relaxed than the somewhat turgid defensiveness of his published works. He sent a postcard of sultry brunette Pola Negri to Konopath, with 'If Only I were Blonde' (a contemporary song) written in mauve pencil on the back. Konopath was not amused. Darré had had these enthusiasms: the sun-worship, the physical culture and nudism, the free love. He understood them, but had outgrown them, forced in part to do so by economic pressures and by his experience with the Baltic farmers. With his military bearing, loyalty and enthusiasm for the cause, and attractive appearance, he became socially popular.[27]

Despite an occasionally condescending attitude, he did not attack his fellow Nordicists directly, but his writings often depicted 'urban romantics, vegetarians and nudists' as being foolishly naïve in not understanding Germany's real problems. He rejected reformism for outright revolutionary change. He found an enemy—liberal, atomised, nomadic capitalism. Before he joined the Nazi Party, he was careful not to attack Jews in his writings. And he found an aim, to attack urban life, and to improve the race through eugenics. He was offering the sensitive an ideal, and the bewildered a clear line of action. Later, Darré was to be written off by the Nazis as a useless theoretician, but during the 1920s he was seen as daring, radical, but above all a practical man of action. Men who see on a long term basis nearly always conflict with those who support or are committed to existing structures. To those geared to making what existed work, Darré was a dreamer, because he called for *total* change; however, his fellow Nordicists were not so committed; they were searching for a new society, and new structures, and here he could offer detailed prescriptions.

Although his career prospects remained bleak, he found increasing fulfilment in writing. Six months after leaving Giessen, he began work on an essay about the link between the Nordic people and the peasant. Hans Günther had sent him a copy of a work by Professor Fritz Kern, also a member of the Nordic Ring. This book, *Artbild und Stammbaum der deutsche Bauern, Racial Types and Genealogy of the German Peasant*, published by Lehmann in 1928, set out to show that the peasants of Germany belonged to the Dinaric race, and that the Nordic inhabitants had formed a non-agricultural, horse-riding, ruling class. Kern's book was copiously illustrated with pictures of hideous, goitrous German peasants, squinting and bracycephalic. According to Kern, Nordics were nomadic by nature, and treasured their horses and their ships. They had originally come from the steppes of eastern Europe, and, far from being *sesshafte Bauern*, were incapable of staying in the same place for long.[28]

This argument deeply offended Darré's own concept of the Nordic, which was of an inherently peasant race, always rejecting urbanism for the rural life, unless forced off their own land by unscrupulous capitalists and landowners. Further, 'nomad' was a code word among the circle around Schultze-Naumburg for the international, cosmopolitan, rootless eastern European Jew. Darré had written a pamphlet on the attitude of different races to the homely pig, in which he argued that the reason why Nordics liked pork and Semites did not was that the pig symbolised a settled, peasant existence; it was not a suitable animal for a nomadic way of life. For Darré, everything bad in history came from the nomadic spirit. Capitalism evolved from the robber tribes of Arabia, and spread to Germany via the Teutonic Knights, who acquired the capitalist spirit from the Arabs in Sicily.

The hierarchical, leadership principle similarly came from the absolutist east. Bolshevism was a cover for nomadic exploitation of settled communities. Concrete buildings and flat roofs, soulless, mechanistic international architecture—everything symbolised by the *Bauhaus* was 'nomadic'. If the Nordic peoples could be shown to be a nomadic aristocracy who ruled in Germany over foreign races or tribes, it would undercut Darré's justification of nation and race, to which a close bond between people and soil was absolutely fundamental.[29]

His original essay grew into a long, unwieldy book, originally entitled *Peasants and Warriors*. He described it to Günther, who answered that he did not fully agree with Darré on the essentially peasant nature of the Nordics, seeing them rather as 'Faustian' wanderers. Nonetheless, he should send his work to Lehmann for publication, as his side of the story needed to be put, and Darré was 'one of the few who really understands how great a role the racial question plays.'[30]

The Lehmann publishing House published *Deutschlands Erneuerung*, which had taken Darré's articles between 1926 and 1928. They also published medical textbooks, and works on racial hygiene and eugenics. Lehmann, generally regarded as one of the most extreme nationalist publishers of the 1920s, was interested in works on the German peasantry. Darré's work, therefore, combining as it did eugenics, Nordicism, and the peasantry, found a welcome home with him. He followed the progress of the book closely, asking Darré to stress eugenic measures, especially the sterilisation of the hereditarily sick, but criticising his outspokenness on sexual matters.[31] Both Lehmann and Günther criticised Darré's style. The first draft of the book was over a thousand pages long, and included pages of long quotations from other writers. Darré assumed that his readers were idiots, and laboriously spelled out every single point. 'Your letters are so interesting', complained Günther to Darré: 'How is it that you write so badly?'[32]

Darré set to work to re-write the book, apparently unabashed by the criticism. The result was a densely textured work, which still hammered home every point. The proof-reading was carried out by Richard Eichenauer, a fellow-Nordicist and university friend of Darré's, who later defended him from attack by patriotic feminists like Sophie Rogge-Börner.[33] Günther then persuaded Darré that he should alter the language, and delete words of foreign origin where possible, a request seconded by Lehmann. This was part of the reaction against non-German cultural phenomena referred to earlier. The 1920s radical right, in its concern for Germany's linguistic and cultural heritage, had developed a self-conscious use of words of German, as opposed to Latin, French or Greek derivation. This affected medical and other scientific writing especially. Darré gave such words as *Sterilisation* and *Kastration*, following them with the German-rooted equivalents, *Unfruchtbarmachung* and *Entmannung*. Later *Konkubine* was replaced by *Kebsweib*; *Mutationen* by *Neubildung*.

Similarly, archaisms were revived, and compound words adopted. *Bauernstand* and *Bauerntum* were words which had no direct modern equivalent. However, it was not a poetic or archaic style, but one which 'mixed erudition, allusion and empirical observation; which used both intellectual sources and popular ones.'[34] Darré liked to start a work with a careful delineation of the position he was going to oppose. He moved slowly and ponderously into attacking position, then would assertively make his point; following up each tangential hypothetical point of opposition as it occurred to him, in long, convoluted sentences.

In the space available, it is impossible to cover all his writings and ideas—especially as Darré was a most prolix writer, producing fifty-six articles between 1925 and 1930, besides his two long books. In the next chapter, the most important elements in his early books and articles will be discussed. Not only is Nazi agricultural policy inexplicable without an understanding of his ideas, but their revolutionary nature needs to be stressed. Most past and present interpretations of Darré, whether they see him as unwordly philosopher or prototype chairman of the Milk Marketing Board, have failed to do so.[35]

CHAPTER THREE

The 'Sign from Thor'

BLOOD AND SOIL

In the Introduction, some of the puzzled interpretations of 'Blood and Soil' have been mentioned. By 1940 it was a cliché which, as Konrad Meyer, head of Himmler's planning office, and Goebbels both wrote, had been ridden to death.[1] It was one of those phrases that so perfectly fitted a concept, a mood of the time, that no definition seemed necessary. And since then, the term has been used to cover a range of Nazi propaganda ideas, Nazi cinema, processions and so on. Use of the phrase is usually confined to discussions of National Socialism, although Isaiah Berlin comments that Disraeli's commitment to 'blood and soil' was more acceptable than the 'insane' ideas of that Nazis.[2] At the Nuremberg Trial, in 1948, the trial judge is reputed to have leaned over and asked Darré's defence lawyer, Dr Hans Merkel, 'Exactly what *is* "Blood and Soil?"' In response, Merkel slammed down photographs of sturdy peasant men behind the plough, and healthy peasant women with their children and said, '*This*, my Lord, this is Blood and Soil.'[3]

But of course, while the glowing images of man and harvest were obviously a part of the idea, it meant more than radiant health and high farm incomes. What it implied most strongly to its supporters at this time was the link between those who held and farmed the land and whose generations of blood, sweat and tears had made the soil part of their being, and their being integral to the soil. It meant to them the unwritten history of Europe, a history unconnected with trade, the banditry of the aristocracy, and the infinite duplicity of church and monarchy. It was the antithesis of the mercantile spirit, and still appeals to some basic instinct as a critique of unrootedness. Certainly, it was not a means of romanticising rural life. National Socialist art stressed the endurance of the peasant rather than his bank balance, and was far from the 'happy, smiling peasant' supposed to typify the Nazi picture, while *Deutsche Agrarpolitik* showed the farmer as an intelligent technician. What was the origin of the phrase and did it change significantly over the decades?[4]

In the early 1920s, August Winnig, the ex-Social Democrat who found the SPD too internationalist, brought the phrase into political prominence. His programme included keeping peasants on the land and improving the physical condition of Germany's working class, a continuation of the great debate over German industrialisation of the late nineteenth century. Georg Kenstler, himself a Transylvanian German who was exiled from the enlarged State of Rumania in the 1920s, meant by 'Blood and Soil', the title of a magazine he founded in 1927, an integral link between the tribe and the land, a link to be defended by blood, if necessary. One pan-European nationalist group of the 1980s proclaimed 'Blood and Soil', to be 'What *we* mean by nation, everything we got from the magic fluid of our ancestors, and from our sacred land: these are the eternal truths the Marxists can never know.' The phrase seems to have acquired more mystical overtones for today's radical nationalists than it had for the German nationalists of the 1920s. Darré took this phrase, and wove it into his own work, which set out to show the absolute primacy of the peasantry.[5]

In his two major works, he defined the German peasantry as a homogeneous racial group of Nordic antecedents, who formed the cultural and racial core of the German nation. Peasant stock provided the source of population for urban growth. Since the Nordic birth-rate was lower than that of other races, the Nordic race was under a long-term threat of extinction. It was also threatened by urban life itself, as Nordics felt ill-at-ease in towns. Since the urban rate of population increase was lower than the rural one, it followed that as the peasant stock drained into the town and there ceased to breed, the peasantry, and hence the core of the nation, would die off. Darré stressed that the peasantry could not be replenished from the townsmen. The urban population had lost its racial and spiritual capacity to become peasants again. This point separates him from most late nineteenth-century land reformers such as Flurscheim and Damaschke, together with others who wanted urban workers to 'return to the land'. 'Blood and Soil', a phrase which in England has somewhat abattoir-like connotations, was intended to express this unity of race and land.[6]

Darré did not see all nations as possessing this characteristic. Other races and groups had different imperatives. He described the Mediterranean peoples, such as the classical Romans, the Spanish and others, as combining a vigorous individuality with a need for a strong state; whereas the essence of the peasant nature was to be anti-state, but to have a strong communal feeling: kin rather than class. He was enthusiastic about the Chinese, who he saw as the other great peasant nation, but Semitic and Tartar peoples were nomadic, and lacked the urge to settle on the land. This meant that they failed to develop habits of thrift, law and self-government, and remained intrinsically opposed to the peasants of other nations, whom they would always exploit if they could.

These kinds of distinctions, these excursions into the true soul of a nation, were also made by other nationalist writers. Spanish nationalists, for example, liked to contrast Mediterranean traditions with northern European ones. They saw the Mediterranean peoples as essentially more *civilised* than the eternally primitive Germanic tribes of the north.[7] For them, of course, civilisation was a good thing, not a dirty word.

Germany, for Darré, was a failed giant. It had been culturally conquered by a series of alien ideas. These included Christianity, the empire of Charlemagne and capitalism, the latter derived from late Roman times. Absolutism and exploitation came to Germany from the Teutonic Knights, and the artificial power-state they established in Prussia. The knights had contracted anti-tribal leadership ideas and an obsession with money from the Arabs in the course of the crusades. Despite his nationalism, Darré refused to glorify traditional German heroes. The peasants had always been the victims, the losers. Since history tends to be written by the victors, and presents the victor's triumph as inevitable, this meant that Darré now opposed many conservative values. The church was a camouflaged army: the mass murder of Saxons at Verden was its bloody crown. Charlemagne had slaughtered thousands of German peasants at Altenesch. 'Not the crucifixion, but this should be our holy place', he noted. The cause of the peasantry's neglect by the German state went back to the 'collapse of the Peasant Wars' in 1525; 'it cannot be blamed only on the Jews'. His support of the peasants in the peasant revolt of 1525 again marks him off as a radical populist from other German *völkisch* writers; after all, the peasants had been ferociously brutal on that occasion. Nomads, capitalists, feudal lords, cardinals, the Roman empire, the Holy Roman Empire, the vulgar plutocracy of the Wilhelmine era, law, church and education—he rejected them all. The spirit of nationalist movements can often be gauged by their *ideal* time in history. This folk-nationalism tends to refer in central Europe to an idealised eighteenth century; in Denmark to the Viking era; in Britain to the early medieval period. Darré went back to the fifth century A.D., to the period before the Roman invasions.[8]

The peasant symbolised a web of values which decayed in the monotony and sensory deprivation of industrialisation. Not only racial factors were at work here. Peasants who left the soil broke their link with the land, and lost their peasant nature. This was due to the importance of cultural characteristics in Darré's definition of Nordicism. Peasant culture was a result of its Nordic origin; but the Nordic idea could not survive urban life. Continuity and kinship were essential to the peasant. If migration to towns broke the generations-old web of belonging and rootedness, the peasant soul was dead.

Many Germans at this period felt that the peasant needed protection against the pressures of urbanism and industrialisation. Darré was practically alone in his criticism of attempts to cope with the problem by returning people to the land. 'A peasant as such can probably be created, but not a Nordic peasant ... Nordic blood can no longer master other blood', he wrote in 1926. 'The task is to maintain the existing [peasantry], not to settle doubtful racial elements'. Darré reserved most of his venom for the German equivalent of well-meaning Fabians, accusing them of looking at the symptoms of the problem rather than its economic, social and racial roots. He saw land reform as a means by which the peasant remnants would be contaminated by urban workers from an ethnic rubbish bin. He argued that capitalism destroyed the peasantry, by splitting farmers from their land, by encouraging a mercantile attitude. Once the peasants had gone, Germany could no longer defend itself, and its downfall was inevitable. National renewal was impossible without the 'life-source', the peasantry. So Darré presented a choice between inevitable doom and a radical, painful effort towards a new society.[9]

Not only did he hammer at a reformist attitude, which merely tinkered with the problem, but at a romantic appreciation of rural life. 'Sentimental and edifying discussions of the evils of modernity, and the superiority of a pure and noble German soul', or 'urban romanticism' were frequent targets.

> Urbanised intellectuals think they can cure the problems with allotments and home ownership, with 'rurban' settlements and homesteads, with vegetarianism and nudism, without noticing capitalism's diabolical sneer at the idea that [these things] ... can make the system healthy again.

How was the 'system' to be rectified? 'If we want to build a truly populist state, then we must build it from the agricultural realm.

Industry and trade will be incorporated into the national economy according to its needs'.[10] This conflation of Nordicism and peasantry, which, as mentioned earlier, was not common among his fellow Nordicists, was the rationale for his proposal to form *Hegehöfe*, or hereditary holdings. Society would divided into corporations, each forming a self-governing chamber, and only farmers from a *Hegehof* would have a vote. The Napoleonic Code sub-division of farms would cease (this only applied to western Germany in any case), and primogeniture would be introduced. Foreclosure and sale of farms would be forbidden. Younger sons would form the militia and the governing class, leading to the gradual formation of a peasant state within a state. Because food producers held the whip hand over parasitical cities, and because of the cultural and racial vitality of the peasant world, it would conquer the town: the *pays réel* would triumph.

The *Hegehof* was the forerunner of the Hereditary Farm Law of September 1933. Although the corporatist element was not included in the legislation, the rest of the original plan was incorporated without much alteration. In a later chapter, the results of the experiment will be examined.

For the survival of the peasant state, protection would be needed, both from competitive capitalist methods of food production, and from malevolent manipulators of the world's commodity markets. An autarkic trade structure would help defend the peasant from destructive mercantilist forces within Germany. Darré proposed a national marketing authority, which would incorporate all primary and secondary food producers (i.e., biscuits as well as milk), and deal with distribution and sale of food. There would be quality controls, but no crop quotas. Here, there were precedents for the idea of taking farming out of the market economy. Eventually, Britain was to establish various marketing boards in the 1930s, and Hugenberg evolved a marketing board in 1932, which some writers have seen as a forerunner—even inspiration—of Darré's, but Darré's plan came from Gustave Ruhland, Professor of Political Economy at Freiburg, who was charged by Bismarck with examining the effects of the depression on European agriculture in 1884. Ruhland became adviser to the Agrarian League of the 1890s. His agitation against commodity trading in the 1890s had actually led to the closing of the futures market on the Berlin Stock Exchange, according to his English translator.[11] His three-volume work, *System of Political Economy* (Berlin, 1908), proposed a controlled marketing system for agriculture, and was a profound influence on Darré, who tried to have his work re-published. He organised study groups among his agricultural economists, to produce proposals on Ruhland's lines for his proposed marketing corporation in 1931-2. The marketing corporation obviously had a potential for political corporatism, and when it was established in October 1933 aroused interest in other European countries, especially France, where corporatist philosophy was especially strong.[12]

Since dissatisfaction with current marketing was so prevalent, Darré was able to attract a good deal of sympathy for his plans. As the National Union of Fertiliser Retailers pointed out bitterly in a telegram to General Schleicher in 1932, no one was interested in the problems of the retailer![13]

If Nordics were equated with a rooted peasantry, so the cultural qualities of the peasantry reflected its Nordicness. Peasant motifs in German art, past and present, were later to be distinguished by Darré in works on peasant culture which he commissioned. *Deutsche Agrarpolitik* and *Odal* carried articles on peasant customs and ways of building, farming.

The rural architecture of the settlement projects so dear to his heart had to reflect peasant building traditions. This was particularly popular with farmers who had complained of the institutionalisation of Weimar architecture, with its flat roofs, leaking metal windows and unsuitable concrete walls. Buildings made of local timber and stone, with roofs steeply pitched against wind and snow, seemed more sensible. The attack on fashionable 1920s architectural modernism, a style which rapidly became entrenched in an era which saw some 75% of all housing put up by public authorities, came from many quarters. Hans Günther attacked 'the housing projects of radical architects' as 'the work of the nomads of the metropolis, who have entirely lost any concept of the homeland', while the Dresden Professor of Architecture saw 'nomadic architecture' as inferior to 'folk architecture'. Nordicists emphasised the Asiatic nature of flat roofs, while farmers found themselves colonised by local government architects who put up Bauhaus type cowsheds. Both objected.[14]

It was an architect, known for his imaginative and radical designs before the First World War, who was to offer Darré hospitality when he became unemployed in late 1929, and who first introduced him to Hitler. Paul Schultze-Naumburg's country house, Burg Saaleck, in Thuringia, was a centre for *völkisch* activists and peasant ideologues, such as Dr Georg Kenstler, the refugee from Transylvania mentioned earlier, former *Artamanen* member, and editor of the magazine *Blood and Soil*. Schultze-Naumburg wrote in 1940 that the *Erbhof*, or hereditary family-sized farm, expressed a divine law of order in the world, and that Blood and Soil, also, was an undying law of nature. The interesting point about him is that he was a successful professional man, leader of the progressives before the war, who began to feel in the mid-1920s that national architectural styles were under attack by a cosmopolitan, heartless, featureless modernism. Frick sacked Gropius from the Bauhaus and made Schultze-Naumburg head of a new architectural university, which concentrated more on crafts and the applied arts. Most of its ex-members emigrated (although one modernist architect, Mies van der Rohe was a signatory to a patriotic appeal by Schultze-Naumburg in 1933). Those who have suffered modernist architecture since, at the whim of the public authorities, may feel that in this one point, at least, the radical *völkisch* men of the 1920s were utterly right.[15]

A pantheistic religious feeling replaced organised religion for Darré and his circle, sectarian differences being subsumed in this attack. A naturistic holism was seen as compatible with the commitment to rationalism and the scientific approach. It is interesting to compare here the pre-First World War pantheistic rationalism of Haeckel, the biologist, and founder of the Monist League. Haeckel wrote in 1905,

> While occupying ourselves with the ideal world in art and poetry ... we persist ... in thinking that the real world, the object of science, can be truly known only by experience and pure reason. Truth and poetry are then united in the perfect harmony of monism.[16]

Darré's call to rescue the peasantry and renew the Nordic race was legitimised for many of his readers by its interweaving with more widely discussed and accepted issues; eugenics, population policy and peasant development. His claim to serious learning marked him off from many *völkisch* writers of the period. Coudenhove-Kalergi, for example, mentioned in chapter Two, prefaced his call for a new nobility with a ringing declaration that mere scientific veracity was irrelevant. Darré and the eugenics movement, however, demanded attention to scientific principles. Lenz, geneticist and racial hygienist, and co-author of a text-book on eugenics, attacked his critics for being unscientific in 'maintaining that biology contains humane principles ... *We* know, on the contrary, that science is value-free.[17] ... We must follow the facts of human heredity wherever they take us', a position very different from that of the *völkisch* Langbehn; 'the final end of science is to deliver value judgments'.[18] Scientific relativism is to be found both in Engels and in the New Left to-day. Despite the constant charge of 'mysticism' levelled against Darré, he believed in the existence of an objective, unalterable truth, and in the possibility of reaching it, if only the state and the 'old order' could be swept away. In *Das Bauerntum*, especially, Darré had aimed at a great work of popular synthesis, like Bölsche's *Love-Life in Nature* which had inspired him in his youth. It was designed for a large readership, and was written, not as political polemic, but with near religious intensity. Despite the two hundred references in this book, he failed to cite Spengler anywhere because, Haushofer argues, Spengler's pessimism and determinism would have undercut his argument.[19] Darré's belief that suitably argued evidence could convince others was untypical of the quasi-existential German Right of his time. Given Moeller's distinction, 'The Left have Reason, the Right have understanding', it is perhaps indicative of the extent to which Darré was totally revolutionary that in this he conformed neither to Left nor to Right: imperfect in both reason and understanding, yet reaching for both with an almost febrile intensity.[20]

The anti-church campaign was as much a part of this rationalist, republican attitude as it was nationalist. It would be quite wrong to imagine that it was merely a propaganda exercise. In fact, from a Nazi point of view, it was counter-productive. But for Darré's small circle, which later included Rosenberg on this issue, the anti-Christian, and pre-Christian ethos was strong. Rosenberg noted in 1934 that 'the SS, together with the peasant leader [Darré] is openly educating its men in the Germanic way, that is, anti-Christian'.[21] Hitler, who frequently called on an unaffiliated, unsectarian God in his speeches, left his public position uncertain, but Darré mentions his surprise at hearing Hitler 'openly describing Christianity for what it is—a religion for *Untermenschen*' in front of a respectable mixed Nazi gathering.[22]

In 1934, Darré called on the NSDAP to declare its full opposition to the clergy, and in 1942 he felt 'a great sense of liberation' when his second wife decided to leave the Church. Native religion was better than imported Christianity, just as native law was better than the Napoleonic Code, or the Civil Code introduced in 1896, which favoured the creditor over the debtor. Darré underwent a religious experience in 1934 at the base of a large standing stone in the *Odenwald*. 'What we experienced at the February Stone was a sign from Thor', he wrote.[23]

The attempt to interfere with peasant Christianity was often unpopular, and pagan calendars sent round to farmers were torn up.

Some of the attempts to instil this fundamentalist approach were simply trivial, in fact comic, as when Darré circulated a request not to call pets, seeds and vegetables by the names of the Norse gods: he attributed it to a 'circle of reactionary colouring'. But of course, if one was reading of an attempt to prevent pets and vegetables being called 'Mohammed' in a Muslim country, it would seem understandable. In Britain, many religious groups would object if a cabbage was called 'Mary Mother of God'. The very fact that it is hard for us today to take Darré's circular seriously indicates the desperate nature of the rearguard action he was fighting, in trying to awaken interest in the old German religion.[24]

Darré continued his link with the Nordic Ring during the 1930s, and, on appeal from Rosenberg, helped to recruit members for the faltering group in 1936. Contact was maintained with the Danish peasant movement, and joint 'Nordic Gatherings', at which Danish gymnasts participated, were held until the outbreak of war. But contacts between the Third Reich and Scandinavian countries gradually declined. A close federation between Scandinavia and Germany was not one of Hitler's major policy aims, whatever the rhetoric, and the Nordic movement found itself squeezed out. Eventually, members operated on the periphery of influence, in danger of harassment by the security forces.[25]

Peasantness was linked with the need for a healthy upbringing, fresh air and exercise, wholemeal bread. It merged oddly with the spirit of the urban health reform movement of the 1920s, attacked by Darré in 1931 as 'romantic'. He started a Peasant University, which stressed physical fitness. He boasted that the puny and undeveloped children of agricultural workers became, after a few weeks, healthy and bright enough to challenge the 'prevailing cliché of the dumb, fat and clumsy peasant'.[26]

But how was the peasant spirit defined? Culture could to some extent be demonstrated by presenting and analysing peasant artefacts and customs that existed. The spirit was more difficult to determine, but to Darré it was everything that was sane, whole, balanced and healthy. 'All that I want is to realise the following perception: Confucius × [times] Lycurgus × Old Rome × Prussianness and the Nordic Idea equals Germanness.[27]

The picturesque nature of this prescription should not obscure the underlying seriousness—and bleakness—of his commitment to the restoration, maintenance and increase of the peasantry. His writings on peasant culture are indeed notable for one omission: they contain little that is sensuous, spiritual or emotional in their appeal, and in this comparisons with English ruralists of the 1920s and 1930s falter. He was concerned with the morale and survival of the peasantry. It was survival or death, for if the peasant went, with him went nation, racial identity and creativity—in short, history itself. The danger was total: defence must be absolute, down to the most minute and particular detail.

A State consists of three things ... a people, their national territory, and a State authority [*Staatsgewalt*]. These three ... possess an internal connection. It is precisely this connection between the people and its territory and political order that gives that order its particular characteristics; it is moulded from the organic form of the people. The character of our State will not be determined by foreign areas, as occurs in the great colonial Empires—nor through foreign nations.[28]

What did Darré envisage as the ideal Nordic peasant state? How true is the allegation that it was essentially imperialistic, that 'Blood and Soil', in Tucholsky's striking phrase, 'started Green but became Bloody-Red'? Later chapters will describe in detail the history of Darré's efforts to oppose imperial expansion in the late 1930s. But this 1935 formula clearly emphasised the close link between land, culture, people and State. It associated nationhood with the mixing of labour with the land and had 'Little Germany' overtones. While return of Germany's lost colonies was never Hitler's primary aim (continental revision of the Versailles Treaty came first), there was a sizeable group in the Nazi leadership, as well as in the more conservative older elements in Germany, who strongly supported it. The Blood and Soil group around Darré became noted for its opposition to colonial adventures, and Darré wrote an introduction to a work called *Why Colonies?* in 1934, making this point forcibly.[29] At this stage it might be useful to start by looking at a key passage, which appeared first in 'Attitude and Task of the Farming Population', *DE*, 1930, was reprinted in 1934, but was deleted in 1940.[30]

The article attacked imperial expansion, but considered various solutions for Germany's (alleged) overcrowding. Not only did Germany not have colonies for settlement, but overseas settlements lost touch with the homeland, 'a colossal stupidity from the national biological standpoint'. Population control was 'a castration morality'. He urged instead the reclamation and resettlement of Germany's old lands in the Baltic and Poland, the 'most obvious place' for a Germany 'already embroiled in the Eastern problem'.

Of course, by omitting these words in 1940, Darré was deliberately passing up an attempt to align himself with the regime, and instead disassociated himself from the actual eastward invasion that had taken place in 1939.

Darré's attitude, though militant, concentrated on the return of previously German territory.

> The German people cannot avoid coming to terms with [the Eastern problem]. The Slavs know what they want—we don't! We look on with dumb resignation while formerly purely German cities—Reval, Riga, Warsaw[sic] and so forth, are lost to our people. *Why shouldn't other German colonial settlements of past centuries—Breslau, Stettin, even Leipzig or Dresden—be next in line? But this would mean submitting to a dangerous error.* The German people cannot avoid a life or death struggle with the advancing East. Our people must prepare for the struggle ... only one solution for us, absolute victory! Furthermore, the concept of Blood and Soil gives us the right to take back as much Eastern land *as is necessary to achieve harmony between the body of our people and geo-political space.* [Darré's underlining].[31]

It was on this rock—to confine German revisionism to previously German land—that Darré finally broke. Outside what he deemed to be German or formerly German territory he wanted no part of territorial expansion. Hitler, however, had attacked any variant of Internal Colonisation as pacifist, urging territorial expansion instead.[32] Although Darré was not a Baltic German, he shared their feelings about the German minorities trapped in the newly independent Baltic States, which, like newly-independent countries everywhere, had turned on their own minorities, and were in the process of expropriating and/or expelling them. During his period in Finland and East Prussia, he had seen something of the problem at first hand. Hitler, as an Austrian, was attuned to a different set of interests, while the conservative nationalists of Weimar Germany still longed for the glories of the old Reich. Edgar Jung, lawyer, and influential author of *Herrschaft der Minderwertigen* a member of Othmar Spann's corporatist and élitist circle, and friend of von Papen, corresponded with Darré at length on the future of the Baltic States vis-à-vis Germany. Jung supported a 'Greater Germany imperial ideal', which would incorporate the Baltic States and various national minorities from Central Europe. He saw the day of the *völkisch* nation as over, and attributed problems in the Baltic States to the problem of 'hooligan nationalism' there, among the Estonians, etc. Darré, however, rejected the power-politics ideal of a Germany ruling the splintered nations of the North.[33]

Given his differences with the neo-conservative intellectuals, and his attacks on Christianity and the German nobility, it may at first sight seem surprising that these Catholic intellectuals and nationalist nobles seemed attracted to, as well as challenged by, his ideas. After all, their conservative nationalism was a world away from his radical racialism.

The sympathy was due to Darré's fusion of biologically racialist ideas with the idealism and voluntarism of Moeller and Spengler, writers who were not themselves biological racialists. His belief in the organic link between peasant and soil went further than mechanistic eugenics. In this, he bridged the mystical fervour of messianic political thought in Weimar Germany, with the mainstream eugenic and agrarian reform of the established political parties. His new contribution was the revolutionary, almost nihilistic radicalism. But it was expressed in a new and striking way: not in the existentialist pronouncements of a Moeller, but with the specific, detailed proposals of the earnest social reformer.[34]

Furthermore, in his eugenics, Darré was appealing to a movement which cut across the party political spectrum. There were social hygienists on Left and Right, besides a fringe of, on the whole, right-wing racial hygienists. Two prominent members of the Social Democratic Party, Alfred Ploetz and Grotjahn, were especially concerned at the genetic implications of the welfare state which they supported. A eugenics policy was needed as 'a corrective to the otherwise inevitable degeneration that would ensue'. The anti-semitic element seems negligible; many geneticists and eugenicists were of Jewish origin. The racial hygiene movement was concerned also with infant health care, abortion, and a variety of problems then considered hereditary, such as alcoholism and tuberculosis.[35]

The marriage certificate was first urged by Agnes Bluhm in 1905. As in Britain, returning armies after the First World War brought back with them a very high rate of syphilis, and in 1920, the National Medical Council debated whether to introduce compulsory health certificates before marriage, this to include an investigation into genetic background. Eventually, marriage advice centres were established in most urban areas, where they offered voluntary counselling on hereditary problems.[36]

Some of Darré's more extreme proposals, which will be discussed later, appear less strange in this context. Since eugenics and racial hygiene are sometimes presented as a specifically German obsession, it is perhaps worth mentioning that America was the first country to introduce compulsory sterilisation for hereditary defects in some States, together with a compulsory venereal disease check before marriage; that Russia and Sweden started Institutes for Eugenic Research in the 1920s, and that to this day there are several European countries (and some US States) where a health certificate before marriage is compulsory.[37]

The winner of the Social Darwinist essay competition of 1900, Schallmeyer, described himself as a 'radical democrat', who wanted to educate society towards a 'progressive eugenic awareness'. Fabians in Britain, including Bernard Shaw and Beatrice Webb, supported eugenic action as a progressive, modern measure. So Darré's intra-racial eugenics was taken as that of a well-meaning, socially concerned, compassionate reformer. He was working within an established social and intellectual framework.[38]

Logically, this meant that conservatives often attacked Darré. For example, in 1939, an *Action Française* writer criticised Darré's racial theories as 'comic to the highest degree' and a 'laboratory experiment'. He considered it a typical example of Germanic solemnity and excess. Lenz, however, the progressive eugenicist, approved Darré's idea of the hereditary farm.[39]

Darré's mixture of cogent polemics and scientific fact was useful propaganda for the committed nationalist movement, and his publisher made every effort to push the works. They appeared on many a *völkisch* reading list, along with *Varuna*, Ludendorff's book on the Jesuits, Maeterlinck, and works on practical astrology.[40] Although for the first two years (1928–30) only 1400 copies were sold, some 150,000 copies were sold within the decade.[41] Darré's work focused on so many current controversies, including the existing debate about the meaning of the concept of race. *Völkisch* writers, with their concentration on anti-semitism, had included cultural, spiritual and religious determinants. The biological-racial Social Darwinists discussed wider racial differentiations—negroes, not Jews. Neo-conservative writers tended to concentrate on the endangered racial élite within a single population pool. Darré's own arguments were defensive, but the defence was against expected attacks on his 'materialistic philosophy', rather than on his unkindness in thinking in racial terms at all. He saw his recommendations for sterilisation of those with non–communicable but hereditary diseases as a kindness, rather than as brutality. He argued that it would enable the victim to enjoy a normal sex life without worries about his children's fate.[42]

His first book had described the German peasantry, the endangered life-source of the nation. In his second work, *A New Nobility From Blood and Soil*, written at Burg Saaleck between December 1929 and March 1930, he described in more detail exactly how the new society would work, and how the new nobility would be formed.

BREEDING A NEW NOBILITY

Racial bodily conformity alone is not enough to mark the State with the spirit of the predominant race, if a spirit alien to the race remains predominant in the State. The German State, the Third Reich which we seek again, cannot be realised just by selection with a predetermined physical shape for sole aim. For this reason, we have to instil the real German ideals of the State into the spirit of future German youth.[43]

In order to accomplish this, Darré turned to English education as a model. He believed that it produced a communal team spirit by cultivating instinct and social feeling. It emphasised character and team sports, the principle of 'government by the governed'. In short, the main function of education was civic maturity, 'if we summarise ... Fichte and Savigny'. German education was founded on 'the Prussian barrack system ... In a barracks it is impossible to create the self-disciplined, self-ruling duties ... It is hierarchically and authoritatively regimented'. But his support for a self-governing community included admiration for the military virtues. He praised the teachings of von Moltke, Scharnhorst and Schlieffen. 'To them is due the salvage of the German state from the hands of the assassins and looters in the years after 1918', while von Seeckt's autobiography gave useful advice for the young;

The essential thing is action. It has three phases; decision, born of thought, preparation and action itself. In all three phases the will directs. The will is born from character. This last is more decisive in action than enthusiasm. Spirit without will is worthless: a will without spirit is dangerous.

Darré concluded, with enthusiastic eclecticism, 'These words of von Seeckt's demonstrate the possibility of uniting the educational system of England and Germany.'[44]

This criticism of German education differed sharply from the generally accepted picture, and indeed, from the admiration for German technical and commercial education, both before the First World War and since.[45] To suggest that German education was too individualistic is almost as heretical as to suggest that Germany lacked ruthless efficiency. To ally the criticism of excessive individualism with a critique of excessive autocracy, appears to compound the confusion. Yet there is a connection. Darré saw the mindlessness of the autocratic hierarchy as encouraging a petty-minded type of competitive striving. Both qualities were alien to the initiative and intuition he wanted to inculcate, and the correct educational process could set free the human will. Indeed, man's capacity to control the environment, added to human control of eugenics, could mean that the solution of all social problems was in sight, given a fully scientific attitude. 'At the beginning of all eugenics is the human will,' he wrote in *Eugenics for the German People* (1931). Darré saw the latest advances in knowledge in the biological sciences as enlarging human freedom by enabling human material to be manipulated. Research should be left to qualified experts, but 'the question, what should one make of the once-and-for-all given hereditary factors, what is needful and what is unnecessary among this genetic pool . . . is in the first place a political matter'.

As explained before, Darré alleged a causal relationship between German 'peasantness' and Germany's national survival and creative capacity. The new peasant nobility, secure on its inalienable farmstead, would replace the existing old, worn-out, despised ruling class. It would also correct the tendency of the 'best blood' to die out through warfare and lower birth-rates. Breeding aims would include bravery, health and intelligence, and concentrate on the peasant, whose tribal attitude to family and children had once been instinctive, but which had been debased through contact with alien, mercantile lifestyles. Now, a better understanding of Mendelian laws would soon enable the laws of heredity to be codified, and the 'genetic inheritance' which had ensured centuries of creative achievement, would be preserved.

This argument owed much to the now little known works of Madison Grant and Lothrop Stoddard, two American writers, who, before and immediately after the First World War, complained that the Anglo-Saxon element in America was becoming submerged by non-Anglo-Saxon immigration, and that intra-racial decline was taking place, the upper classes breeding themselves out, while the 'Jukes' inherited the earth.[46] Darré read of their ideas in Günther's *Racial History of the European Peoples* (London, 1927). However, Darré laid greater stress on the role of the human will, which he fused with the belief in biological determinism. His specific recommendations bore out this voluntarism. He argued that it was impossible to know whether a 'rogue gene' might recur, so any young German who could support a wife should be able to marry: but his wife's heredity must be checked. State intervention should be confined to investigating the character and health of young men, and educating them to choose a wife 'correctly'. Here, the eugenic role of the woman was seen as more important than the man's. The man, the achievement-oriented, voluntarist, creative element, could prove this worth by achievement and status. The woman could be gauged only by her child-bearing potential.[47]

This argument brought abuse on Darré from all sides. Fellow Nordicists were often opposed to purposive breeding. Günther said it was 'a chicken-farm mentality', and it was seen as an attack on family values. Despite Darré's stress on the monogamous nature of Nordic marriage, he was accused of suggesting polygamy. The Young Nordic League called his proposals 'unscientific, immoral, an oriental *haremwissenschaft*.[48] One girl member reproached Rosenberg for allowing Darré, 'a man who wrote such an immoral book', to speak at a meeting on German culture.[49] He managed to annoy men and women alike by picturing the woman as an equal (if separate) partner, the adult, healthy, responsible mother. This ideal, with its emphasis on the natural, and the beauty of womanhood was more attractive to intelligent women than the depraved and salacious sensuality with which Zola and Fontane had presented their passive, sofa-lounging female victims, and was obviously close to the patriotic feminism of late nineteenth-century Germany. But it also brought him unwelcome attentions. He complained in 1934 that he had to cope with every kind of attack, 'the outbursts of hysterical women, who annex the holy idea of breeding to their own lack of erotic restraint.'[50] Not only the more emancipated women attacked him. He shocked a jocular Nazi dinner party, especially the women guests, into total silence when, after the birth of his second child, he announced that if the baby had not been 'all right', he would have had it exposed.[51]

Women were to be classified into four groups: those whose

marriage was to be encouraged, those whose marriage was merely permitted, those who could marry but not have children, and those who must not marry. The latter group included madwomen, prostitutes and recidivist criminals. Natural children would be assessed separately.[52]

Darré stressed his association of the town with miscegenation,

> the danger of uncontrolled introduction of inferior blood with natural children. One thinks of the large towns, where the dark-skinned student, the coloured artist, the Jazz trumpeter, the Chinese sailor, the fruit merchant from Central America, etc. ... feel perfectly at home, and can often leave behind an eternal souvenir.[53]

Again, we find a note of fear, a sense of the nation as helpless victim of the exploiter. What Darré failed to mention here, perhaps because it was too fresh in his readers' minds, was the so-called 'Black Peril', *Schwarze Gefahr*, as the Moroccan and Senegalese troops used by the French occupying force in the Rhineland were called by *völkisch* writers. But he was trying to write moderately and defensively, in order to appeal to as wide an audience as possible, and also to avoid recent and limiting historical references to what was meant to be a scientific programme for social reform. Out, too, went references to von Seeckt's role as Freemason, or victim of Freemasons, comments on the number of top army officers who had Jewish wives, and other outbursts which appeared in his early letters.[54]

Darré's discussion of the mechanism of eugenics implies that he and his circle had considered the problems involved. How could racial 'quality' be defined? Qualities in different races were equal but different. The problem was to understand the limits of the possible. Following Günther, but without Günther's somewhat journalistic concept of the mental qualities of the different races, Darré argued that there were no completely pure races; that interbreeding had gone on for hundreds of years in Europe, but that there were pockets of original peoples here and there, and that Mendelian laws resulted in a patchwork of different genetic distributions, with one racial type dominating another, rather than a homogenised 'melting-pot'. Even if an individual possessed racial purity, 'it does not follow that capacity will effectively manifest itself'. So selection had to continue for generation after generation. By selecting desirable genetic combinations in the children a racial creation would take place. 'The idea of an original race leads to racial chauvinism ... but the idea of eugenics leads to a normal utilitarianism.'[55]

The ideal of this selective breeding was Nordic man, the long-lost, half-forgotten progenitor of Germanness, and the Nordic ideal was ensconced in the peasantry. But the Nordic race could not be created just by exhorting Nordic peasants to have children. Race by itself was not a sufficiently strong factor to determine the form of the state, if what existed was alien to the racial spirit. 'The great currents of world thought circulate too profoundly.' Thus, a revolutionary transformation had to take place in the form of the state before it could be adapted to the intuitive political ideals of the *Volk*. Here, mechanistic eugenic determinism became subordinated to a more integrative social philosophy. 'To the material facts of Race must be added the consciousness that it must have the kind of conditions that are proper to it, so that it can really create itself, so that the earth can be in some way prepared, so that the corn can begin to grow.'[56]

This apparent dichotomy runs through Darré's writings on eugenics, race and the state, and reveals his awareness of factors not amenable to deterministic analysis. He defined the state as a reflection of the customs of a homogeneous tribe; but it must also, because of its capacity for negative and destructive effects on the *Volk*, play a formative role. It was not neutral. The need for identification of people and state stemmed from this definition. Once fixed in form, states tended to be invulnerable to pressure from below. So institutions that did not reflect popular custom and feeling had to be abolished by force.

Since his 1933 legislation attempted to put his major ideas into practice, it is worth looking at Darré's actual method of racial selection, once he was given the opportunity. It does seem to demonstrate that a coercive racial selection was not envisaged, and lends emphasis to the distinction, drawn earlier, between a defensive intra-racial eugenics, which aimed to prevent the disappearance of a group, and the expansionist super-stud mentality popularly associated with Nazis. Under the Hereditary Farm Law, only farmers of German and 'similar' stock, who could prove descent back to 1800, could inherit the protected farm. To the annoyance of many radical Nazis living in border areas, this definition included Polish farmers, who were also enraged at finding their farms rendered not only inalienable but, at one blow, unsaleable. The nationalist press in Warsaw took up their cause.[57] The SS marriage order, implemented in 1931 by Himmler with Darré's help, made SS approval for its marriages a condition of membership, 'solely on grounds of race, or genetic health'. Darré had to put some pressure on Himmler to get the marriage consent regulation through, and had wanted a Scandinavian SS to be formed at the same time, to make it easier for Germans to marry Scandinavians—but without success. Racial education was part of the curriculum of the peasant university at Burg Nauhaus and the SS Racial Office (part of the SS Race and Settlement Main Office). Examples of this work include a circular sent to the SS education department, suggesting that a textbook be produced showing photographs of good racial stock. Darré offered a textbook on horses as an example. SS leaders were shown films on Blood and Soil, harvesting, ploughing, to persuade them of the desirable nature of the life.[58]

The key point, though, is the voluntary nature of these activities. Darré did not try to enforce compulsory breeding laws. He did not incite riots against Poles and Jews and demand the compulsory sterilisation of the unfit. In fact, he circulated his staff ordering them to stop boycotts of Jewish shops in 1935.[59] He looked to racial education to create what he called a 'positive racial consciousness', rather in the way in which today, especially in the USA but to some extent in England, television and other media make special efforts to present the black minority in a favourable light in drama series, children's programmes, and so on, Darré wanted farmers and their families to be educated into racial consciousness—White is Beautiful—as part of a process of instilling a sense of identity. It was seen as a rescue operation for a vanishing breed. Even the SS marriage laws could be breached without serious sanction, since dismissal from the SS (a volunteer group) was not in the 1930s the end of the world.

Once a breeding pool of Nordic peasants had been achieved, leaders would emerge, chosen on grounds of ability. They would form a Chamber of Nobles. This 'nobility by recognition' was Darré's attempt to overcome what he saw as a problem in meritocracy. He thought that in the long-run, meritocracy damaged society, by draining the lower classes of their more able individuals. It failed, too, to ensure the continuation of the valuable genetic inheritance of individuals who proved their worth. But a *principle* of merit would underlie the hereditary nobility. The brave would be acclaimed by their peers, chosen to fight and lead. The ideal nobility was unprivileged, dutiful, hard-working and worthy of its status: Darré referred to Plato's Guardians here. The nobility would be intimately connected with the ruled. To avoid deracination, the nobility would remain part of the peasantry, and lead a farmer's life. Only in this way could the inner intuition necessary to rule be maintained, and caste distinctions avoided. Such distinctions 'conserve, they do not create'.[60]

This criticism tellingly displays his lack of conservative orthodoxy, which traditionally seeks order and continuity, as opposed to spontaneity and radical reassessment. His criticism of capitalist meritocracy clearly had something in common with that of Weber and Pareto, and recognised the problem of a non-egalitarian society, where classes lose their natural leaders. But his attempt to combine stability, justice and flexibility in his ideal society was drawn more from a private vision of an ideal, tribal society (though one, apparently, shared by the Nuremberg trial judge, who criticised Darré for not realising that true democracy was invented by primitive German tribes, 'long before there were any Jews there'). The identification of cultural and racial factors formed a self-correcting circular system: if a Nordic race could be re-created, it would govern according to old Nordic ways, providing that existing institutions were abolished. Peasant intuition, wisdom and the mutual recognition accorded by independent freemen, would ensure harmony and good sense.

This ideal was not Utopian, a description that implies the argument that, if the laws of nature could be abolished or transcended, man would be better or happier. The argument was rather that, if man lived according to the laws of nature, becoming aware intuitively of his links with the world around him, he would be better and happier. The attempt to escape from these bonds could only fail, and lead to a cycle of industrial growth, collapse and misery.

The vision of the Nordic peasant soul portrayed by Darré was ambiguous. The peasant was strong, the source of all productiveness and creativity. But he was also fragile; his soul could not survive cities, capitalism or industrialisation. Darré was driven by his fear for the Nordic peasant, yet his political support was attracted by his projection of its strength. The defensive, racial element later became lost in ministerial practice, in exchange for pursuit of peasant productive potential.

During Darré's early years in office as a minister, close friends and backers absorbed his view of the peasant as inherently vital; the source of national health, fertility and creativity. This eventually mutated into the belief that the creative, entrepreneurial smallholder could contribute to, and co-exist with, industrial society. His belief in the sensitivity of the Nordic soul, its endangered quality, was abandoned by others, including some of his followers.

In any case, a serious contradiction remained in his fusion of spiritual and racial elements. He argued that peasant cultural continuity, the living tradition, was lost in urbanisation. But he also argued that peasantness was genetically inherent in the Nordic people. If this was so, then there was no reason why 'sound Nordic stock' could not be rescued from the towns and returned back to the land. By arguing that peasantness could not re-emerge from submergence in town life, he subordinated his racialist argument to his concept of spiritual peasantness, and, by inference, wrote off the 70% urban population of Germany. This was not an 'unworldly' tangent to his philosophy, but an essential part of it. Whether in fact he would have carried out his beliefs is open to considerable doubt. Just as, when it came to the crunch, he was unable to bring himself to exclude peasants of Jewish ancestry from the Hereditary Farm Law, it seems most unlikely that he would have excluded all townsmen in practice.

It is probably apparent from the foregoing discussion, that Darré was a strange mixture of visionary and realist. He used his selected scientific arguments, learned references and agricultural knowledge in pursuit of a goal whose nature was extremely radical, even in those days of ferment. He saw the re-ruralisation of Germany as the only viable alternative to inevitable national destruction. The next chapter will discuss how he came to think, mistakenly, that Hitler's movement towards the peasant vote in 1930 was his opportunity to carry out his aims. Darré was, indeed, to storm his way to an invincible position in agrarian circles, through the optimism, passion and energy produced by this conviction, and was to obtain ministerial office. Hitler found Darré a useful theorist and organiser for a period of crisis, but when Darré kept faith with his own vision, he was, like many other revolutionary ideologues, discarded.

CHAPTER FOUR

The Conspiracy

Blood and Soil is regarded today as one of the fundamental elements of Nazi ideology, and Darré is seen as a standard National Socialist ideologue. How did he come to enter politics, and how was it that he became an activist in the NSDAP? In many ways, Darré does not fit the *Reichsleiter* type. He joined the Party late, in May 1930; he was from a once wealthy, upper middle-class background, and he had no background as a party political activist, despite his brief membership of the DVFP in 1923.

He had become acquainted during the late 1920s with National Socialist activists and sympathisers, most of them living in Thuringia, and interested in two things; peasant-agrarian problems and the decline of German culture. He knew Theo Habicht, and Habicht was Gauleiter of Wiesbaden in 1928 before going to Vienna. Habicht also knew Himmler, and was a Party speaker. Darré kept in touch with him after his first meeting with Hitler. Darré stayed with Schultze-Naumburg on and off, between 1928 and 1930.[1] As mentioned earlier, Schultze-Naumburg was a respected architect, who had been a leader of the progressive movement before 1914. His country house was a centre of party activity as well as a meeting place for Nordic Ring sympathisers. There, Darré met Hans Holfelder, a NSDAP member who infiltrated the *Artamanen* group in 1927, and died violently in 1929 and to whose memory he dedicated *New Nobility*. He also met various neo-conservative intellectuals there. He knew Hans Johst, playwright and author of a popular work on Schlageter; met Jünger and, later, Heidegger. He was befriended in 1929 by Dr Hans Severus Ziegler, editor of the Thuringian NSDAP *Gau* newspaper. Ziegler, one of the radical peasant ideologues of the circle, admired Darré's work, and wanted him to work for him as a journalist and political organiser. Darré also met Günther at Saaleck, in January 1929 after three years of correspondence. Konopath was a frequent visitor, and joined the Party shortly after Darré.[2]

Despite these contacts, Darré did not join the NSDAP until he was offered a post within it, and his approach seems to have been almost subversive. Despite his strongly nationalist feelings in the early 1920s,

Darré's main interest by this stage was the plight of the peasantry. He had tried to found a 'Union of Noble German Peasants' in 1928, with Dr Horst Rechenbach, later to liaise between Darré and the SS Race and Settlement Main Office; but it foundered in the agricultural depression.[3] Darré now got together with Kenstler, editor of *Blood and Soil*, to organise and finance a 'nationwide network of cells among radically-minded peasants', and is supposed to have asked Ziegler to finance the idea, but had first thought of approaching Hugenberg, the head of the German Conservative Nationalist Party, for finance and help for a political agrarian organisation: his publisher, Lehmann, persuaded him that it would be a waste of time, and suggested that Hitler might be more interested in the idea.[4] This suggests that Darré was looking for a party to carry out his programme, a political machine to infiltrate for his own views, rather than entering the NSDAP as a committed loyalist.

In 1930, the NSDAP agricultural manifesto appeared, designed to appeal to the farming vote. The authorship of the manifesto is not known for certain—indeed, for decades it was assumed that Darré had written it, but according to Erwin Metzner, Keeper of the Seal to the National Peasant Council, it was produced by Konstantin Hierl, head of the NSDAP Department II (Labour) and Himmler, though Gregor Strasser, Feder, Kenstler himself, and Werner Willikens have also been put forward as possible authors.[5] In many ways it foreshadowed later Nazi policy, calling for hereditary, inalienable farms, and debt relief, but there were references to the need for unity between town and country, the desirability of industrial development in the east German provinces, which Darré and his circle could not have approved. The manifesto was openly a party political document, and emphasised national and infrastructural economic aims as well as agricultural reforms. When Darré first read the manifesto, he was interested enough to copy it down word for word.[6] Despite their reservations, he and Kenstler were struck by the document's opportune appearance, and decided that it offered potential for their own plans. There is an undated document in the archives of the NSDAP Agricultural Office, obviously by Darré, which was written for Hitler some time between 1930 and 1933, and which discusses the possibility of a peasant coup, to be carried out by force against Germany's big cities. This may have been Darré's first offering to Hitler in 1930.[7]

The correspondence between Darré and Kenstler was couched in conspiratorial terms, using hints and initials. On April 12th, 1930, Darré wrote,

What I wanted to say is this: if at all possible, I would like to spend the Easter holiday at Wiesbaden with my wife; however, would make [that] dependent on the state of our affairs; if, for example, H. Munich [Adolf Hitler] should come to Weimar during this period, obviously I don't go. Concerning the agricultural programme of the NSDAP, I should not express an opinion on it at this preliminary stage for tactical reasons. Perhaps we will be able to broach the subject in a different way, which I think would be more sensible in view of the different situation. In any case, I could discuss the ... details with you verbally as soon as possible.[8]

The Schultze-Naumburgs were acquainted with Hitler through the Bruckmanns, the Munich publisher. First Frau Bruckmann, then the architect, told him about their clever young friend, with his knowledge of farming and organisational capacities. The idea was now to find Darré a post in Thuringia, so that he could work with Ziegler and Schultze-Naumburg. Darré wrote to his wife on 27th March 1930 that Ziegler and Kenstler planned an agricultural policy institute, under his leadership, to be financed by the NSDAP, and that direct negotiations were under way with Hitler.[9] While his was angling for a meeting with Hitler, he learnt that the lack of anti-semitism in his two books had attracted unfavourable comment from Hitler.

Today I received a long letter from Sch-N. who was at Bruckmanns [the publisher had social-political 'evenings', where Hitler would meet Bavarian society] together with A.H. on Wednesday evening. Result: A.H. had virtually no idea what was intended for me. He did know my name, but not my *Bauerntum*, which he had only heard about. A lively debate developed between A.H. and Sch.N. which, however, was fruitless, insofar as A.H. was not properly aware of what I was after, and Sch.-N, was not the most suitable representative, because he lacked knowledge of basic agricultural principles. Also, A.H. had been falsely informed over my *Bauerntum*, insofar as he believed that I didn't sufficiently interpret the Jewish problem in terms of their parasitical essence. Briefly and clearly, you can see from these few indications how things stand.[10]

Darré went on to say that Hitler was clearly interested in meeting him personally.

On 7th May 1930, he wrote to Ziegler that 'Frau Sch-N.' was making arrangements to catch Hitler 'for a couple of hours or for a weekend' through Frick. At this stage, his publisher, Lehmann, wrote offering to pay him 600 marks a month if he would work for the NSDAP in Munich as their agricultural organiser. The letter still exists, though ironically enough Darré's account of his recruitment to the NSDAP, with his salary paid by outside sources, was not believed by his U.S. Army interrogator, in 1945.[11]

Ziegler was dismayed at the prospect of losing Darré to the Brown House. A letter to Hess, refusing the post, was drafted, but not sent after a phone call from Hess himself.[12] Darré took up his activity for the Nazi Party in June 1930, his salary paid for by his publisher, and possibly other private sources. One writer suggests that a small Munich industrialist, Pietzsch, also financed Darré's salary.[13] Lehmann's intervention over the salary suggests a certain ambiguity in Darré's position. Although Lehmann, who died in 1935, was the only neo-conservative publisher who continued unhampered by the Nazis after 1933, the fact that he virtually bought Darré's way into the NSDAP gives some idea of the extent to which Darré must have been seen as an outsider.[14] The idea that Darré was soft on the Jews did not help, and he never became one of the favoured inner circle around Hitler. His party number was high, around 250,000. He was one of the very few late party members to achieve high office.

Darré took up his post in June 1930. He moved to Munich with Charly, whom he married in 1932. Saaleck was still a centre for what was now avowedly Nazi activity, and in mid-June he returned there for a conference with Goebbels, Frick, Goering, Rosenberg, von Schirach, Günther and Konopath. Konopath became a member of the Race and Culture department of the NSDAP Dept. II.[15] Eventually, co-operation languished between him and Darré, who discarded first Schultze-Naumburg then Konopath, as his interests shifted more and more to agricultural problems, and away from the Nordic movement. 'Keep Paulchen out of Party disputes, he would only get them back to front', he wrote, of the ageing architect.[16] He wrote to Nordic Ring members apologising for delayed letters, and saying that agricultural problems left him no time for the Nordic Ring. Konopath sank into obscurity, and in 1932 his magazine, *Die Sonne*, was taken over by the Skald, a secret *völkisch* order of elitist, revolutionary principles, which will be discussed in the next chapter.[17]

He missed Saaleck, but there were compensations in Munich, such as evenings spent with the Bruckmanns, and Darré remarked with awe that 'In this very house Houston Stewart Chamberlain stayed ...' an indication as to how strongly Chamberlain was admired among the National Socialists.[18]

His contacts with Hitler in the Brown House in Munich were few, and he was obviously in awe of him. 'Even I, as his adviser, can scarcely get to see him,' he noted, and told inquirers who wanted to contact Hitler to first make contact with Frau Bruckmann instead.[19] One letter to his ex-wife was written on notepaper headed with an imprinted picture of Hitler, presumably standard issue for top Party members. Other letter-headings included pictures of the Deutsche Eck in Coblenz, and the Sans-Souci orangerie in Potsdam. He referred to the 'atmosphere of the leader, of leadership' in another letter explaining that he was too busy to keep up his private correspondence.[20] These letters, though, seldom included comments on politics. When they did, they had that lecturing, didactic and public tone which Darré was prone to adopt with his first wife. He told her one anecdote about Hitler and Hindenburg which sheds some light on the NSDAP reaction to Hindenburg's refusal to appoint Hitler as Chancellor after the 37.4% Nazi vote in July 1932. Hitler had been sent damaging material about Hindenburg, which he had refused to use. The result was that 'the Jesuit Brüning' gave Hindenburg 'dictatorial powers' [when Hindenburg became Chancellor] and Hitler's head could be 'on the block' at any moment. The lesson he drew was that ruthlessness was necessary in politics.[21]

Most of his letters concerned his daughter, Didi, of whom he was very fond, and complaints about money worries (in September 1932, *Reichsleiter* salaries were halved, and payment was often delayed).[22] Darré, despite his formal, inflexible bearing, loved children, and they always responded to his sensitive and perceptive approach to them. One visit from Didi had to be put off because of an election campaign in north Germany and there were problems with others, because of Darré's occasionally homeless state. He moved from furnished flat to furnished flat, occasionally staying with friends, while paying half his salary over to Alma (a situation which continued until she married again). Despite financial problems, and the embarrassment of acquiring a new, pretty and younger wife with no home to take her to, he felt relief at the separation. He asked Lehmann for money to help pay for the divorce. Lehmann, who admired Alma's strength of character, was shocked at the separation, but Darré insisted to him that he could not have carried on his political work with her, and needed to break free.[23]

Darré continued his weekly letters to his ex-wife until he learned that his daughter had contracted tuberculosis and needed expensive hospital treatment. He wrote an unpleasant letter to Alma, accusing her of neglecting Didi, and demanding that she pay the medical costs. While it is always difficult to judge ex-marital quarrels from the outside, his attitude here seems appalling, and understandably enough, Alma broke off the 'deep and warm friendship' that Darré had vowed in mid-June, 1930. Oddly enough, he seemed proud of the letter, and kept several copies of it in his files, although it clearly showed the frantic anger of a man under stress.[24]

THE BROWN HOUSE 1931–1933

Given Darré's interest in corporatism, one would have expected some degree of sympathy between him and other early Nazi theorists, such as Feder and Otto Strasser. However, he regarded Feder as too urban-oriented, and Strasser as an incipient Bolshevik. One of his early diary entries refers to the 'rather Communistic impression' made by Strasser's black Russian smock and red shirt.[25] Strasser's corporatism, however, did stress 'Germanic self-determination' rather than the 'Fascist satrapy', and Darré later found more in common with Gregor Strasser, head of the Party's Agricultural Organisation until Darré took over.[26] Strasser told Lehmann that Darré was 'doing first class work'. Darré described Strasser as 'with the right ideas, very clever, but generally not easy to handle ... Once you win him over, he's a powerful ally'. The relationship became more difficult as a row developed between Strasser and Hitler, which ended in Strasser's resignation.[27]

He made a good impression on men like Otto Wagener, ex-General Staff Captain, corporatist and follower of Feder, who Himmler told of Darré's 'wide knowledge of the world' and agricultural expertise.[28] Darré's *Neuadel*, more readable and polemical than the earlier *Das Bauerntum*, was a good introduction to these Nazi circles. His views on land division and settlement appealed to the 'urban' Nazis because of their implicit hostility to conservatives and land-owners. The young Heinz Haushofer, son of General Karl Haushofer, was taken to meet Darré and Hess in Munich in late 1930, and was struck by Darré's ideas and interest in the now-neglected Gustav Ruhland.[29] The agrarian adviser to the Pan-German League, von Herzberg, visited Lehmann and asked to see Darré, and Class, author of pre-war nationalist best-seller, *If I were Emperor*, head of the

Pan-German League, also wanted to meet him. Ernst Hanfstaengl, script-writer with UFA and old friend of Hitler's, insisted on reading *New Nobility* before visiting England, in case he was asked about it by the English press. Altogether, Darré had become a major figure in radical nationalist circles.[30]

But what lay behind Hitler's sudden need for an agricultural expert? Why did he think the German farmer was ready to vote for the Nazis? How committed was Darré to the Nazi Party, and why did he think his plans for a peasant rebellion belonged there? How true was Darré's later claim that his usefulness to Hitler gave him more independence than the other *Reichsleiters*?

German agriculture is widely seen as split into two extremes, the small peasant farm, and the Junker landowner. This picture is exaggerated, as the typical Junker farm was about 1000 Ha. in size, and much of that would be barely usable. In short, the Junker 'latifundia' were more the size of a substantial tenant farm in England than the Scottish landowner's 20,000 acres, and changed hands more often, showing greater social mobility than England's landed élite, according to a recent study.[31] The apparently inefficient small peasant farm had shown surprising growth potential between 1880 and 1925, indeed, at a time of agricultural depression, this sector had grown more than any other.[32] In many ways, the small farm was more resistant to depression, economic controls and inflation than the heavily mortgaged large farm. In 1914, some 30% of Germany's population still lived on the land. The east Elbian Junkers usually farmed their farms themselves, and in many ways had more in common with the small farmer than either had with the large estates of the south-west, often the remnants of Holy Roman Empire holdings. However, the Junkers had considerable political power. They presented themselves as the backbone of the nation, and attracted subsidies and other protective measures which sometimes affected small farmers badly. Political divisions, as represented in the farming organisations, were minimal, and essentially, both groups wanted protection from marketing problems. One potential source of political divergence lay in the fact that peasant areas of Germany, such as the Rhineland and Bavaria, had a large Catholic population, and Catholic peasants tended to vote for the Centre Party, while larger farmers voted for the DNVP conservatives.

There was one substantial group that fitted into neither category, the small-medium farmers of Lower Saxony and Schleswig-Holstein. Here, the land was fertile and intensively farmed, and productivity per hectare in Lower Saxony was higher than anywhere else. The inflation of 1923 was more damaging to this category than to the heavily mortgaged large farm, whose debts were in some cases wiped out. The boom of 1925, fuelled by short-term American investment, encouraged the most advanced farmers in north and north-west Germany to expand. Dairying and market gardening thrived on the heavy black soils of German Frisia. The 'Red Earth' settlement associations of Westphalia bought up land at high prices for smallholdings. But these areas were trapped by the Depression of 1928 which pre-dated the crash of 1929. High levels of debt combined with a drop in food prices and very high taxation levied by the Social Democratic Prussian parliament to pay for urban workers' social welfare. Of course, all German agriculture was affected, but the farmers of north and north-west Germany, traditionally liberal voters, felt especially penalised. They had been exhorted to invest and expand, and were now going bankrupt, where 'dog and stick' farmers had kept their heads down and kept their farms.[33]

By early 1928 nearly half Germany's farms were making a loss. Average profit levels had dropped to a mere eight marks per hectare, while taxes and other government burdens amounted to an average 26 marks per hectare. In some cases, interest rates were 10% (a real interest rate of $c.7\%$).[34] The result was a spate of bankruptcies and foreclosures. Large farms were reduced rather than affected, because in an emergency it was possible for them to sell off some land, though there was deep psychological resistance to splitting up estates.[35]

This agricultural emergency resulted in sporadic armed resistance on the part of the farmers, which nearly led to civil war. Farmers bombed tax offices, and shot off rifles at foreclosure auctions. These 'protesters' were often the best educated and most widely travelled. Unlike the peasants who had exploded in eastern Europe in the 1920s, they found an ideology ready and waiting in neo-conservative radicalism. Prussian intellectuals and ex-*Freikorps* activists like von Salomon poured in to help, with the impassioned patriotism of the beleaguered minority, rather than the confidence of the powerful. Unlike the East Prussian grain growers, the north German small farmers were outsiders; they failed to fit the numerous interest group structures, and had the worst of all worlds as a result of the existing system of taxation and subsidies. In English terms, they were more like the Northern Irish, or Anglo-Welsh farmers, than like the East Anglian arable farmers, who had always had the ear of the Conservative Party. The black flag of revolt was raised in Schleswig-Holstein, its symbol a plough and a sword. One popular song of the movement began,

I put a bomb in the tax office,
and dynamite in the county parliament,

phrases which are catchier in German.[36] The refrain of the song called
to Hitler, Ludendorff, Erhardt and Hugenberg for aid, a more eclectic
group, in party-political terms, than might at first appear.

National Socialist attacks on finance capitalism and Jewish
speculation were willingly accepted by men who saw creditors, banks
and international commodity markets as the root of their problems.
The Narodnik streak in the German Youth Movement supported the
attacks on tax officials and bureaucrats. The ruling Weimar coalition
saw nothing to be gained by subsidising these traditionally liberal men,
while the DNVP was more concerned with subsidies for powerful
agricultural producer groups. The protest movement was crushed by
heavy prison sentences in 1930. Ironically, this destruction of violent
and anarchic action left the way open to the infiltration and
propaganda devised by Darré. Now revolutionary action had failed, the
farmers turned to a legitimate party that took their grievances
seriously. The emotional Anabaptism which had underlain 1848 almost
as much as the Peasants' War of 1525 welcomed the outspoken attacks
on large landowners, banks and freemasons. Darré was not a brilliant
orator, but he offered the promise of restitution to the disinherited, and
future security. His speeches stressed the national importance of the *Bauer*,
and the programme carefully worked out by him and his staff over the
the next two years corresponded to the perceived problems of the time.

Hitler's need for the rural vote was based on a shortfall in the 1930
vote, although this was sharply increased from the mere 2.6% of the
vote in the 1928 national election to 18.3%. Hitler realised he would
have to increase the party's attraction to new groups. The Protestant
small farmers were an obvious choice, disaffected, nationalist,
radicalised, and split geographically between support for the
conservatives and for the Liberal Party. Unlike the Catholic farmers,
they had no party devoted to their own interests. There was a
remarkable disparity in Nazi support between Catholic and Protestant
small farmers, 73.9% in Protestant areas, to 12.3% in Catholic areas.[37]
This was due, not to sectarian appeals in or from the party, but to the
fact that the largely Catholic Centre Party already occupied Nazi
ground in many ways. It was devoted to peasant interests, anti-big
business, anti-big trade unions, covertly anti-semitic, anti-Communist,
and had strong regional loyalties. The Nazi Party was a peasant party
in Protestant areas, but a proletarian party in Catholic areas.

In June 1930, Darré organised a sub-division of the NSDAP Department 2 (Labour) into the Agricultural Organisation, a quasi-independent network of activists, and sympathisers, whose task was not only to infiltrate the farmers and peasants unions by electoral means but to discuss land reform, agricultural improvement, structural changes in tenure and other issues. He kept their activity moving through a daily stream of memoranda and orders. Most of these memoranda still survive, and offer a detailed picture of the way in which National Socialists organised their recruitment and propaganda activity.[38] Unlike many other departments, Darré's agricultural department in these early days did not waste time and energy in in-fighting. His leadership was accepted from the start, and it was recognised that few others could have shown the same combination of expert knowledge, nationalist reliability, organisation and hard work. One post-war author refers to the 'exceptional organisational and tactical abilities' of the 'National Socialist cadres'[39] He made a good impression on his fellow Nazis at this time, especially those below the rank of *Reichsleiter*, precisely because he was not an 'old fighter' but an outsider. He came from a higher social background than many old Nazis, and this again helped him in the first few years, when non party political support was especially welcome to men who had been virtually full-time political activists since the war. Such activists, aware of the isolated nature of their lives, always see new blood as a breakthrough, a sign that they are beginning to win over 'the others'. Darré belonged to a traditionally liberal family background, the wealthy merchant with overseas connections, and so represented a particularly unusual type of recruit. *Hitler Regiert*, a hagiographic portrayal of the *Reichsleitung* written in 1933, laid great stress on Darré's elegant appearance, his good looks, his well cut suit and polished manners, to an extent which shows how useful he was to the party's image.[40]

Darré did not always reciprocate, and in the early days in the Brown House, some enmities were instant and mutual. 'Goebbels hates me,' he noted in his diary. 'It must be because of his black blood', a double edged crack at Goebbels' dark colouring and his Catholicism.[41] Within a few months, Princess zur Lippe was complaining that Goebbels had given widespread publicity to a pamphlet attacking 'the racial work of our friends' [i.e. Darré], expressing not only personal animosity, but a deep dislike of Blood and Soil concepts. Darré told Konopath that Goebbels did not understand 'scientific racial concepts', and hoped to use him in the Propaganda department to combat

Goebbels' influence. Darré managed to win control of the AD journal *Landpost* from Goebbels' propaganda department. He also warned Konopath about Himmler. 'Many people laugh at him, but his influence over Hitler is greater than people realise.' [42]

The Gauleiters, mostly old party members, were jealous of the independent nature of the Agricultural Department (or AD), and Darré's direct responsibility to Hitler. Darré had made this independence a precondition of his work. He was in a strong position, not only because the party needed his combination of expertise and organisational ability, but because Hitler had taken the initiative in approaching and recruiting him. In January 1934, the *Reichsnährstand*, or National Food Estate, the new marketing corporation, was again formally promised independence from the Party and direct responsibility to Hitler. [43] Darré established an area network that paralleled that of the NSDAP. One report on the structure of the Agricultural Department in Saxony, demonstrates the similarity. [44] There was an expert adviser at Gau level, and 34 at Kreis level. 1100 NSDAP farmer members were active at farm and village level. Twenty-two party members were representatives in the Chamber of Agriculture. There were 40 speakers who specialised on farming and party matters, while Saxony AD had sent four members to the State and National Parliament. From the Gauleitung point of view, much of this effort could have been better used in strictly Party matters. In the early 1930s, the situation was accepted because it was recognised that Hitler needed to capture the rural vote if the NSDAP was to win an election. But by January 1933 the party was in a stronger position, and the Gauleiter of Saxony ordered the Saxony AD to reduce its strength by 33%, allegedly to save money, but in fact to stave off AD competition. Riecke, a Gau adviser on agriculture (i.e., a member of the NSDAP organisation), who was later to be second-in-command under Herbert Backe, Darré's successor, wrote dismissively in his memoirs of Darré's 'Blood and Soil efforts ... The Westphalian peasants, among whom I worked, had quite different worries'. [45] Riecke was dismissed as agricultural adviser after 1933, and worked for the volunteer labour force thereafter.

Where the AD broke new ground, and began to approximate to the planned farmer's corporation, was in its system of expert advisers. There were nine such experts in Saxony alone, each with an area of expertise, such as market gardening, agricultural co-operatives, settlement, and fowls. They were accompanied by seven more advisers on general peasant affairs, such as education, debt and appropriation problems, and insurance. Darré insisted that all agricultural experts had to be farmers. Party members were trained in special courses laid on by the AD. In one such series, five local areas trained 700 members on how to become elected to positions in the agricultural unions, such as the *Landbund*, how to recruit peasants in small villages, how to speak at public meetings and how to write letters on political subjects to the local newspapers.[46] The intensely regional nature of Germany's political life and press meant that each area needed to produce its own propaganda, quickly, flexibly and effectively, so that local issues could be used where necessary. The Saxony report estimated that some 18,000 to 19,000 letters, cards and pamphlets were produced between January 1932 and January 1933, while 50 directives were sent round internally. All areas were canvassed at least once, even the most remote, and meetings were well attended by peasants, usually between 130 to 400. Darré quickly became a popular speaker, not so much for rhetoric and bonhomie as for content. The German farmer was a political animal, already organised into various farmers' unions, and producer co-operatives, and it was easier to reach a politicised group than it would have been to start from scratch. In December 1931, Werner Willikens, one of the AD's leading members, was elected president of the *Landbund*, a major success for the AD.[47]

In July 1932, Hitler's Party attracted the largest vote it was to have before gaining power, 37.4% of the vote. In a system of multiple minority parties, it was an overwhelming victory. The north German Protestant farmers and village and small towns had voted for Hitler—averaging some 78.8%. In some areas in the Geest, Nazi votes were 80–100% of the total.[48] The smaller the village, the larger the proportion. As a reward, Hitler told Darré he might be offered the Prussian Agricultural Ministry if the NSDAP came to power. At this stage, Darré was not expecting a position in the national government.[49]

He insisted that special attention be paid to the agricultural co-operatives, especially the Raiffeisen ones, because they were completely independent of state finance. One AD member reported in February 1933 that over 40,000 co-operatives were represented in the Raiffeisen group. 'With a few exceptions, the whole farming population is organised in co-operatives or societies and is bound to one or more.'[50] The implications of this fact for the proposed new marketing corporation were considerable. It meant that much of the basic groundwork was already done, that farmers were psychologically attuned to protection against market fluctuations, and to the idea of cutting out the marketing middleman: further, that a pool of experienced administrators already existed.

AD activists were helped in their organising by the fact that there was general agreement about the failure of the market economy in agriculture. The only question was how should it be reformed. The activists went as far as opposing leasehold tenure. Instead of seeing tenancies as a means by which the young farmer could make a start, and the smallholder extend or diminish his holding according to market conditions, they opposed the renting of land as a capitalist phenomenon, an unwanted intrusion of the mercantile world. This, if Darré's plans were to be implemented, had serious implications for the land market. In 1925, 12.6% of all agricultural land in Germany was rented, and 67% of farms between two and five hectare.[51] The quickest and simplest way for farmers to expand was to increase the amount of land rented on a full- or part-time basis, but the agricultural depression of the late 1920s reduced the proportion of rented farmland, and in 1933, only ten per cent of farmland was rented. The development of leasehold tenure came under attack from a range of political groups, although the Weimar Constitution of 1919 had granted tenant farmers considerable security of tenure. Both the SPD and the DNVP objected to cost-effective and profit-oriented farming in principle. For the SPD land reformers, followers of Henry George, leasehold tenure represented the exploitation of the peasantry. They supported land nationalisation, or at the very least, the nationalisation of rent.[52] For the German conservatives, it increased the danger of splitting up the large estates, and loss of control over land use. The concept of the peasant as a profit-making businessman, able to buy and sell his land at will, was opposed to all Darré and his staff stood for. They attacked the concept as an inheritance of nineteenth-century liberalism. According to them, it took an atomised, individualistic view of the farmer, who became merely part of a mercantile ethic. For the Nazi agrarian radicals, the small farm was the core of the nation, crucial to its physical, moral, cultural and racial health. As the source of the nation's vitality, the small farmer could not be left to suffer under market forces.

Darré also attracted intellectual support from those who supported the peasantry for quite other reasons. One school of thought saw small farms as more productive, in terms of net deliveries per hectare, than large ones.[53] There was resentment against subsidised competition from large landowners, and especially the unpopular *Osthilfe*, a special subsidy to east Elbian grain growers. The peasant was doubly the victim, paying taxes to support the subsidies, and having to pay more for his fodder. The AD under Darré reviled large land-owners as 'bacon-tariff patriots', and he gleefully circulated their counter-attacks, accusing him of being worse than the KPD.[54]

The AD's attack on leasehold tenure was accompanied by the demand to settle farmers on inalienable medium-sized farms. The campaign to win over the peasant and small farmer was helped by Darré's opposition to the large landowner. Darré's friend Theo Habicht published *Deutsche Latifundia* in 1928, a book which attacked large landholdings, and Darré used his figures in his 1930 *New Nobility*.[55] The AD was united in support of a programme which would divide up the great estates. This did not attract as much opposition as one might expect from large landowners, since many were demoralised and uncertain of their future, while others supported the idea of peasant settlement on nationalist grounds. As long ago as the late 1890s, the chairman of the Agrarian League had called for state intervention—if necessary, by nationalisation—to prevent the dismemberment of estates in private hands.[56] However, at this time, some of the large farmers began to realise the implications of the state aid they had been requesting (and receiving) for decades, and began to move towards a stronger view of property rights.[57] The German conservative party saw Darré as a dangerous radical, which of course, he was, especially to their interests, and any attacks by landowners were proudly circulated to the members of the AD. One anti-Nazi landowner, Prince zu Löwenstein, in an article called 'The Radicalised Village', claimed that thirteen million small farmers had been won over to Hitler. 'A precondition of winning the fight against National Socialism is to ... ensure that Communism doesn't fill the vacuum left by Right-radicalism.'[58]

There were signs that, in their desperation, farm-workers in East Prussia *were* voting for the Communist Party, and Darré, after commissioning a report on the economic situation of state domaines in East Prussia, sent in a team to win the tenant farmers over from the German Nationalist Party.[59] While General Schleicher, the 'Socialist General' who was shot in 1934, was flinging subsidies around in wild abandon to the Bavarian egg producers, the grain growers of East Prussia, and the dairy operatives of Lower Westphalia, telegrams began to land on his desk about the fact that farmers were oscillating between the KPD and the NSDAP.[60]

During this period, arguments about peasant productivity had developed controversial political overtones. Support for the large landowners was linked to the DNVP, the heavy industrial sector, and an *ad hoc* interventionist line on the part of Brüning on agricultural subsidies. This bloc called for the retention of high grain prices, and even talked of the 'duties of the consumer towards national production'. However, the more dynamic, export-led section of German industry, always more sympathetic to laissez-faire ideas, favoured a 'modernised and capable peasant agriculture', which should be incorporated into a more competitive industry. Market responsiveness and varied cropping were the ideals; Holland and Denmark the exemplars. One later writer for the RNS, and editor of the reports of the International Conference on Agricultural Science, the highly respected agrarian economist Constantin von Dietze, wrote a report for the Chamber of German Industry in 1930, which emphasised the virtues of market oriented peasant farming. His co-author, Karl Brandt, emigrated to America in 1933-4, where he wrote articles which, among other things, criticised the Junkers for their backwardness, obstinacy, and continued political power. Darré's background of overseas trading expertise gave him natural allies among many trade-oriented and technologically aware businessmen.[61]

Darré's hitherto unused abilities came into play during these three years. He moved with absolute certainty, laying down a clear ideological line that appealed to many of his members, but also responding to their ideas. His nose for corruption and intrigue kept his members on their toes. His analyses and predictions of motives and behaviour were often proved accurate, excepting only the power of financial incentive, where, as with all Nazis, he grossly underrated its efficacy in increasing production. Darré's group was more exclusive than the NSDAP, which had formed a loose alliance with the DNVP, the Stahlhelm and others, at Harzburg (the so-called 'Harzburg Front'). His farmers were not to belong to the Stahlhelm, because the leader was a Freemason. The press and cinema empire controlled by Hugenberg was a cheat, pushing an anodyne, 'bourgeois' line. *Die Tat*, the nationalist intellectual journal controlled by Eugen Diedrichs, also had masonic connections. About the only group outside the AD where cross-membership was encouraged, was the *Allgemeine SS*, in 1931 a group of a few thousand part-time volunteers. Darré recommended his members to visit their local SS groups, to see if they could be useful, a condescending attitude that seems comically inappropriate in retrospect, but which reflects the relative status of the SS at that time.[62]

Darré's main ally now was Heinrich Himmler. The Race and

Settlement Office of the SS was established in 1932, largely through Darré's efforts, according to his diary, and his affidavit at the Nuremberg trial. (The function and scope of this office will be discussed at greater length in a larger chapter.) In the affidavit, he described his motivation for founding the Racial Office as being to try to instil in German society a sense of fair play and public-school spirit that, he claimed, was taken for granted in English society, at all social levels.[63] However, odd this may sound, it is clear that Darré sensed a lack of co-operation and community spirit in German society. Given his own experiences, that is not surprising. The emphasis on this *Volksgemeinschaft* does suggest that it was deficient to begin with. It has by now become a historical truism that the National Socialist government was structurally chaotic, and consisted largely of factional in-fighting. While this can be exaggerated, Germany in the 1930s did seem better at building motorways and motor cars than at handling its own dissensions—quite apart from internal political opposition. It is one of the many ironies of Darré's life that he envisaged the SS racial élite as a means of bringing the British public school ethic to Germany.

Darré and Himmler were both interested in the agrarian question. Himmler was a trained farmer, like Darré, but he had had no practical experience, and relied on Darré for factual information, and it is probable that Darré, seven years older than Himmler, had influenced the latter in this area. Both men were interested in peasant settlement, and Himmler became liaison officer between the NSDAP and the *Artamanen* in 1930.[64] However, Himmler was interested in Indian philosophy and the occult, and had several Indian friends and contacts in the 1920s and 1930s, partly because he was drawn to these Aryan Nazi sympathisers, but partly because of the political importance of the Indian pro-German nationalists. Too much can be made of the importance of bizarre cultism in Himmler's activities—he is supposed to have sent a party of SS men to Tibet in order to search for Shangri-La, an expedition which is more likely to have had straightforward espionage as its purpose—but it did exist, and was one of the reasons behind the split between Himmler and Darré that took place in the late 1930s. All the same, Herbert Backe thought that Himmler's growing romanticism and mystic interests was a result of Darré's bad influence, and while Darré was not interested in the occult, which he had jeered at in the Nordic Ring, he was capable of inspiring others with his own enthusiasms, and many of his ideas re-emerged through Himmler in a perverted form in later years.[65]

CHAPTER FIVE

The Nazi Minister

In June 1933, Darré was appointed Reichsbauernführer, National Peasant Leader, and also Minister of Food and Agriculture. In September 1933 he presided over the creation of the *Reichsnährstand* or RNS, translated here as the National Food Estate; the agricultural marketing corporation. By 1936, Darré's authority had waned, and between 1939 and 1942, the year he was demoted, it was effectively nullified. The story of Darré's years in office is deeply interwoven with that of his successor in 1942, Herbert Backe. In 1936, Backe was nominated to represent agriculture on the Council for the Second Four Year Plan, and from being second-in-command in the Ministry of Agriculture, he came to be preferred to Darré. He was seen as a more reliable, straightforward administrator, and an efficient and loyal technocrat. From 1939, he was given the title of 'Leader' of the ministry.

The Ministry of Agriculture rose in terms of its share of financial resources from eighth largest ministry in 1933 to the fourth largest in 1944. Its budget rose seven times between 1934 and 1939, compared with an 'average increase of 170% for all ministries'.[1] But it lost its political power. Backe, Darré's successor, only held on to his tenuous position by supporting Hitler's policies, while Darré remained loyal to his principles and lost power. The two men had worked closely together between 1931 and 1934. They pushed the Hereditary Farm Law and the Marketing Law through a largely conservative and hostile cabinet in 1933, and fought to keep agricultural settlement as the preserve of the National Food Estate, when the Ministry of Economics called for more 'urban settlements'.[2] But by 1936, mutual hostility was so great that Backe was appealing to Goering and Himmler to remove Darré, while Darré was issuing a stream of defensive memoranda and letters.

The reason for this shift in authority is crucial to a history of agrarian ideology under the National Socialists. Was there a significant ideological dimension to this feud, which still persists among followers and friends of Darré and Backe? Or were revolutionary political activists particularly prone to such squabbling—in that the habit of intensive opposition to the existing political system was hard to break, and had to be directed somewhere? Was the Ministry for Agriculture particularly affected, or was it symptomatic of Hitler's method of internal party rule, symptomatic of the multiplicity of competing offices established after 1933? The relationship between Darré and Backe does seem to illuminate certain problems of Nazi ideology, style and policy.

One interesting side-issue is what exactly Backe's ideas were. He was, and is, often referred to as a *non*-ideological figure, a practical man with no interest in cranky ideas about Blood and Soil. Goebbels, Kehrl, Speer and many modern authors have taken this line.[3] However, Backe, a trained economist as well as a farmer, had a firm commitment to a planned economy which is hard to reconcile with the picture of the pure pragmatist. He was one of the young Nazi intellectuals, a friend of *Die Tat* writers Giselher Wirsing and Ferdinand Fried Zimmerman. He was also accused of having 'fantastic goals',[4] because of his attitude to de-collectivisation in Russia, but this was due to his fear of losing a harvest during a changeover of methods—a 'pragmatic' preference which was unsuccessful in this case, and, it will be argued, typically so.

This chapter will argue that Backe's motivation, ideas and style as a human being sheds important light on the nature of National Socialism. His belief in planning, action and need for loyalty to a leader led him to join a secret, revolutionary group in the early 1920s. Several members of this group were to hold high office in the Third Reich, among them Theo Gross of the Racial Office. Backe, however, has remained a shadowy figure, perhaps because he was not prosecuted in the IMT Trial at Nuremberg. Even his name has suffered distortion into Wilhelm and Ernest Backe, instead of Herbert Backe, by Speer and Robert Koehl.[5] He committed suicide in 1947, after months of solitary confinement, and his name did not figure prominently in the 1949 Wilhelmstrasse Trial which included Darré as defendant (Case XI). But Backe was nominated by Hitler to the Dönitz cabinet of April 1945, and was respected and trusted by Heydrich. Many of his peers saw him as a prominent and representative intellectual. While agriculture and the peasantry played an important role in his ideas, he placed national politics first, and his loyalties gradually came to clash with Darré's, the more so as Backe's admiration for Hitler grew, and Darré's admiration, such as it was, waned.

I realise that my tension and nervousness are a result of my development being distorted—hindered and destroyed: my hatred of the authors of this destruction [Russia] came about as a result of that.[6]

This uncharacteristically introspective analysis by Backe indicates that his 'distorted' development is worth examining. Such an admission is rare in the autobiographical literature of NSDAP members, and indicates something of Backe's uncompromising honesty. In his prison jottings of 1946, he laid down his own version, in which each phase of his life was shown to give rise to certain thought processes, and an examination of the documentary evidence from 1918 to 1945 bears out his analysis quite closely.[7]

Backe, like so many other National Socialists, began as a member of a beleaguered minority. As a German national, born in Batum in 1896, he entered the Tiflis State Gymnasium in 1905, at a time of increasing and sometimes violent Russian nationalism. His father, a Prussian merchant, committed suicide while Backe was about fourteen years old, and Backe had to work his way through secondary school. Russia's rapid modernisation between 1900 and 1914 was accommpanied by a fierce hostility to all national minorities, and was followed by wartime persecution of many minority groups—Germans and Jews especially. Backe spent the war interned in a camp for civilian aliens, in the Urals, while his elder brother fought in the German army. He was exchanged to Germany in 1918 via Sweden, on condition that he was not conscripted.

The attempt at forcible Russianisation produced fear and contempt in Backe: the hardships of internment, resentment. His first-hand experience of civil war atrocities meant that his hatred extended to the Bolsheviks, although he always kept an admiration for Russia's potential for achievement. His family had lost everything during the war, and were penniless. Backe now had to keep three young sisters, a sick mother, and an elder brother studying engineering. He worked for six months in factories in the Rhineland, while taking his *Abitur*. When Backe lost his factory job in the post-war depression, the family moved to Hanover, where he worked as a drainage labourer in the nearby moorland. When the girls were old enough, and physically strong enough to find work, Backe trained for his agricultural diploma, again working his way through university. He began to study at Göttingen in 1920-1. Vacations were spent working for board and lodging only as a farm agent, as Darré was to spend months doing in 1922. He took his diploma in 1923, with good marks. In many ways Backe was typical of the uprooted and dispossessed 'nouveau pauvre' of early 1920s Germany. He commented in his 1946 *Grosser Bericht* that 'lest this situation of an uprooted family appears particularly striking, in fact there are millions of Germans in a similar situation'.[8]

His first experience of Germany, therefore, was one of defeat and bitterness. He felt 'deep disappointment' at the behaviour of the Germans, which was also a typical reaction for the returnee. He had hoped to find a community at least united in defeat, but instead saw 'egoism and materialistic strife ... the middle classes fighting among themselves,' and contemplated emigration.[9] This sense of disillusionment and mutual rejection in a collapsing 'homeland' helped push Backe into radical and militant attitudes in his early political activities. He refused to join a political party during the twenties, because he believed in a supra-party communal spirit. He clearly had a deep need to identify with a group, to operate within a closed community. His calls for sacrifice in the service of the nation, his emphasis on the value of self-denying behaviour, expressed the need to enlarge the self within a greater whole which is so common in the political drive and rhetoric of countries where the Protestant confessions are a prominent cultural feature. It is interesting to note that on the two occasions when this strong drive was denied in Backe, in 1918 and in 1946-7, his reaction was a self-punitive one: a desire to emigrate in 1918—and hence removal from the desired object, Germany—and thoughts of suicide in 1945, culminating in an actual suicide in 1947. Certainly, the draft testament he left behind is ample evidence that his suicide was motivated not by a sense of guilt or disillusionment with National Socialism, but with a world that he felt had betrayed his ideas.

Backe's first employers in Hanover were all active Nazis, but he seems not to have thought of joining the NSDAP himself until he met Dr Ludolf Haase, then a medical student in Göttingen, in 1922.[10] Backe agreed to join the SA, but still refused to join a Party; and Haase paid his subscription in secret. Backe had resented his inability to fight in the First World War, when his elder brother served in the German army, and he now threw himself into marching and bill-posting. But his role was not confined to strong-man: he soon became the most admired brain and ideological mentor for the Göttingen circle of activists. Haase, as a schoolboy before the First World War, had formed an anti-semitic group in Hanover, and continued distributing leaflets in 1918-22.[11] Several members of this small group in a quiet agricultural area were later to hold high party office, and Haase later wrote, 'When one of our members ... founded the racial policy office of the NSDAP, this was no accident'. Backe, von Gronow, Meinberg and Gross, all held top posts in the Third Reich.[12]

In the last chapter, the takeover of *Die Sonne* by the Skald Order in 1932 was mentioned. Haase was a founder member of this secret society, which was banned by the Nazis after 1933 because of its allegedly masonic nature. An undated secret report on the Skald written around 1936 mentions Haase and Gross, but not Backe, who was also a member, according to his widow, and Princess zur Lippe.[13] Heydrich was later to investigate Backe because of the Skald link; he concluded, shortly before his assassination, that Backe was 'all right'. When Haase became a civil servant in 1942, Hitler had enquiries made, but on learning that Haase worked under Backe, he told Heydrich that there was no need to worry: Backe's loyalty was legendary. Haase is rumoured to have approached Backe in 1933, when Backe became State Secretary under Darré, to pursue the aims of his secret society, to be told by Backe, 'I am not a rebel'.[14]

Haase was galvanised by the radical, militant and anti-semitic rhetoric of the Munich Nazis in 1923, and formed a branch in Lower Saxony, which concentrated on 'good organisation and efficient personal leadership at all levels.'[15] He aimed to bring together 'a group of really *völkisch* men ... round whom the masses would crystalise' when the right moment came. Backe was one of these leaders. He was the chief expert on Russia and racial policy. 'Not all our ideology came from books', wrote Haase; for among their sources of live information was 'PG Backe, who had a special knowledge of all eastern racial questions'. Backe supplied the group with 'a knowledge of Russia, its men, its ... history, and its geography.'[16] The group believed that Soviet rule in Russia could not last, and that in the welter of successor states that would arise from a dismembered Russia, Germany could seize land in the east, and German farmers migrate there in force. This was bluntly stated in 1923, in a speech calling for

> a new man as dictator to unite Germany, throw out the Poles and Czechs, prevent France from hindering Germany's rise. Social legislation, currency reform, and a new land law ... Our colonial future, however, lies in the east, where a new land beckons when Jewish rule in Russia should collapse.[17]

Backe's support for this aim was useful for the Göttingen radicals. The nationalists were split on the question of Russia. As mentioned in chapter One, conservative nationalists as well as social democrats saw Germany's only possible future in a pact with the other 'rogue' state, Russia; but Hitler's criticism of the Treaty of Rapallo was that the Russians should 'shake off their tormentors' before a Treaty could be signed.[18] Backe took a *realpolitik* view rather than an idealistic one. Despite his contempt for the Russians, he recognised Russia's potential in terms of raw materials and human resources. He sarcastically attacked the idea of a Russian–German war in 1925, when many nationalists supported the idea:

> Who will put élite troops against Russia? France or England? No. It will be German troops who will be used to pull the chestnuts out of the fire. A war against the Soviets cannot be lightly undertaken. It would become the whole Russian nation engaging in a national fight for freedom against Western capital ... He should not become the enemy of the Russian people. Germany's future lies in the east, economically, as well as politically. Currently, we cannot have close links with Soviet Russia, but that is unavoidable.[19]

Backe appears to have taken a radical stance on other matters, including anti-semitism and revolutionary methodology, although here again, caution should be used in using Haase where he is the only source of information. He supported the idea of a revolutionary élite, drawing heavily on Lenin's *What is to be Done?* He rejected parliamentary tactics aimed at achieving power legally within the Weimar system. Whether or not to go for parliamentary power was a matter for considerable dissent within the early Nazi Party up to Strasser's resignation in 1932, but Backe was not only hostile to 'entryism', but opposed to the Nazi belief in a mass movement. In his notes for a speech in 1923, he wrote of his preference for a minority élite who would form the leadership of a secret revolutionary group to further the revolutionary cause. The élite would of course represent the true feelings of the masses, as opposed to Bolshevik élitism, where the minority had organised itself to oppose the will of the masses. Leaders of a truly *völkisch* movement would eventually incorporate the masses through a process of 'organic development'. Typically, Backe went out of his way to praise those Soviet leaders whom he considered to be competent organisers and planners, as opposed to the more populist, intellectual and Jewish end of the spectrum. 'Lenin, Rykov, Chicherin, Krylenko, Dzherzhinsky, are worthy of admiration. Unworthy: Trotsky, Kamenev, Zinoviev, Lunacharsky,' etc.[20]

Why was the Russian question so central to the ideology of the northern Nazi radicals in 1923-5? They considered the Greater Germany dream, the old nationalist aim of uniting all German-speaking peoples, as the 'purest pacifism'. They demanded a ban on emigration to America, the incorporation of Flanders, Holland and Luxembourg in a common customs and currency union, and colonisation of the Russian borderlands. Backe, took pains to distinguish this plan from imperial adventurism, 'no solution for the victors or the conquered,' and presumably welcomed German settlers being accepted into Russia as benevolent improvers of the quality of life, and of the quality of *Volk*. Backe and Haase both advocated a strict eugenic policy, not to create a superior élite so much as to purge the nation of those suffering from hereditary defects. Their attitude to Jews seems to have been tougher than that of Feder, Strasser, and other more urban-oriented Nazis. Although quoting approvingly the Eckhart-inspired lines in the 1925 Party Manifesto, 'We will fight the Judaeo-materialistic spirit within and outside ourselves', Haase and Backe also warned that Jews would not be accorded the usual rights of foreign nationals within Germany, and that any alliance with Jews in other countries would be prevented, 'and any return made impossible once and for all'.[21]

Backe's strong will and polemical abilities, his tough attitude and comradeship, were prized in Göttingen, and he was a great loss to the group when he dropped all political activity and moved to Hanover to work on a doctorate in agriculture. Haase commented in his memoirs that Backe had been especially 'important during the early time of struggle', not only because of his sharp thinking and uncompromising stance, but because of 'his ability to create a rounded *Weltanschauung*' and his 'characteristic toughness, which so many Germans lacked.'[22]

At Hanover, Backe became research assistant to Professor Obst at the Technische Hochschule. Obst, who had worked with General Professor Karl Haushofer a year earlier, was a geo-politician whose book *England, Europe and the World* helped to stimulate Backe's views on autarky and Germany's need for a European trading zone which he expounded in his thesis. A plan to visit Russia as agricultural adviser for the Hanover Chamber of Agriculture fell through when the Russian Government vetoed Backe's name: in 1927, his doctoral thesis was rejected on the grounds that it was a work of political science rather than agriculture. Backe suspected political bias against him, and the political implications of the work were considerable, as Backe argued in his conclusion that the colonisation of Russia by Germany was a geo-political inevitability. When Obst was obliged against his will to dismiss Backe, he decided to ignore politics and concentrate on farming. After another year spent working for board and lodging only, Backe married the daughter of a Silesian industrialist, who had worked as an assistant to Obst, and who was herself a radical nationalist, who corresponded with Spengler, and knew the Bruckmanns. She strongly supported agrarian radicalism. His father-in-law provided enough capital to take on the tenancy of a neglected state-owned farm near the Harz mountains, where Backe was able to put his ideas into practice, using small-scale farm machinery, and consolidating and improving what had been a seven-field system. Although the rent was low, the venture coincided with the onset of the Depression, and the collapse in food prices.[23]

Between 1929 and 1931, when his first meeting with Hitler took place, he rejoined the SA. His wife recounts that he did this because his farm labourers were SA men, and he felt it his duty to join them, an interesting sidelight on noblesse oblige. Backe used to go with his men to nearby Braunschweig, where the forbidden brown shirt could be donned, and a march proceed. In 1931, he stood as candidate for the NSDAP in the election to the Hanover Chamber of Agriculture, but lost in the second ballot to a candidate from the *Landbund*. In July, 1931, Backe had an article on Germany's trade and agricultural problems published in the National-Socialist *Zeitung*. Werner Willikens, a friend of Backe, already working with the NSDAP AD, showed Darré the article, who wrote enthusiastically to Backe, asking him to write for the AD. Despite the 'strong impression' that Hitler had made on Backe, he refused to become involved, giving as reasons his need to work on the farm, the economic crisis and his own lack of literary talent. Darré asked him to edit Ruhland's works, and again Backe, who had not read the *System of Political Economy*, although he had enthusiastically quoted Ruhland in 1923, refused, through lack of time. His wife, who had read Darré's two books with fervent admiration, wanted Backe to join the AD and become more active in the NSDAP. The two men exchanged attacks on the profit motive in agriculture ('not reconcilable with the peasantry') and jeered at recent English legislation on smallholdings.[24]

Backe had failed to formulate exact agricultural policies when he met Darré in 1931. The fact that Darré had formulated such policies, and was offering concrete solutions, meant that Backe could throw his energies into implementing Darré's plan without being aware of any possible conflict of views. Here again, Backe's need for a complete, rounded view of a given project was an important factor. His goals were immediate. Well rounded, well thought-out processes had to be followed by potent action and the necessary will to carry it through. In this sense, the two men complemented each other in the first few years of their joint activities. Backe's failure to formulate policy was not because Backe had lost interest in agricultural policy during his years as a practising farmer, as the article published in 1931 which attracted Darré's attention developed his belief that the world trade system was in a state of collapse, and with it agricultural productivity, in so far as it was geared to international markets: it was more a failure to consider detail.

On his wife's insistence, Backe read Darré's two books, and wrote in a second article that Darré's suggested restructuring of German society around the small farmer was the only answer to the 'social problem'. Darré was impressed by Backe's points, laid out with a compulsive air of certainty, and asked Backe to visit a Weimar conference in October 1931, to instruct the agricultural department personnel in macro-economic matters.[25]

Backe refused the invitation. He felt little interest in ideas about self-governing corporations that were being examined at this period. Similarly, although he approved Ruhland's criticism of the international commodity system, ('so absolutely classic ... banks and finance capital above all were the gainers from these events'), Ruhland's proposal for agricultural syndicates made no appeal. Darré later decided (quite wrongly) that Backe's lack of enthusiasm in the invitation meant that his later interest in the AD was spurious, and he wrote after the Second World War that Backe was infiltrated into the NSDAP by Ludolf Haase and the Skald. Indeed, he referred during the war to the 'Hanover clique', whenever Haase or Backe appeared near him in Berlin. This was a source of continual agitation to Backe, because of the investigation by Heydrich.

It was only after hearing Hitler speak to the SA at Braunschweig in the autumn of 1931 that Backe decided to visit Munich, and see what was happening there among the National Socialist agrarian activists.

In the first flush of Darré's enthusiasm and energy, the slight frisson over Ruhland was discounted. He was recruiting whatever talent he could find, to work in the AD and its nationwide branches. Nothing could have been more generous than his attitude to Backe, his respect for Backe's polemical abilities, his encouragement and advice, both political and personal, and his acknowledgement of Backe's role in the early rise to power. In fact, if anything, he exaggerated Backe's abilities. Backe had a gift for stating a situation in terms which appeared to be cogent and analytical, but which, if scrutinised closely, turned out to consist of sweeping statements of principle, without much backing in argument or logic. His need for a superficially orderly thought-system reflected not 'pragmatism' but the need for a highly structured framework of ideas which would provide the certainties he needed. Nearly every memorandum and letter he wrote contained the words *Klarheit* and *grundsätzlich*; it hardly needs an elaborate exegetical or psychological analysis to guess that this virtual obsession with the need for mental clarity and a firm basis expressed its lack in Backe's make-up. Because of this, he could be influenced by potential leader figures. First Darré, then Hitler, held his loyalty in the NSDAP. When Darré's divergence from the National Socialist path became clear in the late 1930s, Backe had no difficulty in deciding which to choose.

His wholly misleading vein of certainty appeared when he finally visited Munich in January 1932. He heard Hitler address the assembled Gauleiters, and the *Gau* agricultural advisers, many of whom Backe already knew from his old SA days. Hitler's speech stressed the importance of the agricultural sector to the German nation. 'Our highest aim is not only to win the city workers ... the NSDAP is not only a party of the cities, but is today the greatest peasant party,' a speech which impressed Backe, for a revealing reason: 'Hitler unified the entire political problem into one simple line'.[26]

Backe addressed the *Landesgruppenführer*, the agricultural cadre leaders, at the same conference, at Darré's request, and spoke about world agriculture. In the discussion that followed. Backe was distressed at the gap between the well-formulated, unified ideology presented by Hitler, and the 'lack of clarity' of Darré and his co-workers. This criticism was not made specifically, as he simply accused Darré's group of ignoring 'agriculture's real problems and the necessary measures'. But a look at his specific criticism of the agricultural department shows that what he was attacking was the lack of a single, basic *ideological* attitude. 'An all-embracing analysis of the forces active in economics was lacking—a precondition for alleviating conditions on the land.' Instead of concentrating their ideas on the collapse of world economic relationships, he was disgusted to find the LBF's discussing the *Genossenschaften*, the *Landbund* 'and other organisations, such as the

Artamanen movement, tariffs, the importance of rent'. Backe's reaction reveals the inadequacies of his yearning for an integral , holistic approach. The tactics adopted by the NSDAP AD were all-important to the eventual takeover of power. Each of these apparently trivial 'single issues' had to be carefully considered to see whether it was absorbable by the NSDAP or an irritant, and tactically the method paid off.

In order to play his part in Darré's programme of infiltrating the Agricultural Unions, Backe stood as National Socialist representative for election to the Prussian Agricultural Chamber, and became a member in April 1932. He disliked this episode. Contact with parliamentary methods reinforced his earlier contempt for them. 'German parliamentarianism will *never* carry out the necessary economic reforms,' he wrote.[27]

During 1933, the two men co-operated closely. The existence of a common Conservative enemy in the shape of Hugenberg, Minister of Economics and Agriculture, from January 1933 to July, when Darré took over, and von Rohr, Darré's most trenchant Conservative critic and Hugenberg's State Secretary, enabled Backe and Darré to make common cause in the fight to keep the AD going, and take over from Hugenberg. The Conservatives still had a majority in the Cabinet at this stage. Early in January 1933, Darré sent a telegram to Backe asking him to help lay the AD proposals before Hitler, in view of the worsening agricultural situation. Although agriculture was beginning to pull itself out of the long depression, few realised this at the time, and prices were rock bottom. The request gives an indication of Darré's lack of confidence in his own abilities to persuade Hitler, his reliance on Backe, but also his sense of security about his own position. Three weeks before the Nazi takeover of power in March 1933, the two men worked closely on the report, which was given to Hitler verbally at a meeting. Backe did most of the talking, as the most persuasive of Darré's staff, and Hitler approved a statement calling for the need for a 'total change of the agricultural marketing and production structure.' The AD staff sent a postcard to Mrs Backe, which reads: 'Dear Ursula, everything went as it should, Hbt. Your husband did very well, Willikens. The report was wonderful, Darré.'[28]

Darré's single-mindedness in pursuit of the enemy was now turned against Hugenberg, whom he described as a 'blockhead', and von Rohr. He sent open letters to Chancellor Schleicher, using a polemical, reproachful, tough tone. Hugenberg rapidly passed a debt interest reduction law, which kept interest rates for farmers below two per cent, but this was not enough to protect his position. He then, although not previously known for anti-semitism, made a strongly-worded speech on foreign policy and racial matters in London, during the summer of 1933, possibly in an attempt to maintain his position in the face of Nazi attacks. However, the conservatives, shocked at the speech, asked Hitler for Hugenberg's resignation, which Hitler was happy to grant, as it removed a powerful member of the DNVP from the cabinet. Von Rohr was another threat, typifying the conservative opposition to the proposed agrarian legislation.[29]

The planned new marketing system, for example, was forcefully opposed by him, because von Rohr realised that the end of Brüning's draconian deflationary measures should see an upturn in economic confidence; and that agriculture should share in the upturn. He feared that to introduce rigidity into the agricultural system by limiting mobility of land use, restricting credit and controlling prices would prevent peasants reaping the expected benefit. Another area of dispute was the AD plan to divide up large estates and settle peasants on them. Hugenberg, especially, opposed this. A legislative basis for this proposition already existed in the Weimar Constitution, which permitted the seizure of land without compensation 'in the interests of the State and the community'. In order to protect himself, Hugenberg went some way to meet the Nazi radicals. He installed a *Reichskommissar* to oversee dairy production, a measure of which Backe approved, as the 'first attempt at controlled marketing', and a 'success'. Hugenberg pressed for his debt legislation to be retrospective. While this attempt to defuse hostility between himself and the radical Nazis was proceeding, von Rohr also kept a low profile, writing enthusiastic speeches about Blood and Soil with no apparent qualms, during the spring of 1933, but attacking Darré and his policies during cabinet meetings.[30]

However the AD, led by Darré, were determined to 'break into the ministry', knowing that their survival as a group of well trained, highly motivated agricultural experts and planners depended on it. The NSDAP was waiting for them to put a foot wrong; their relative independence was feared and envied. Hugenberg, leader of the DNVP, and a millionaire in his own right, owner of a chain of newspapers and film companies, seemed a dangerous enemy. Darré looked for support among the peasant community. 'Send your former leaders to the devil. What have they ever done for you? Suffering and hunger, hunger and suffering,'[31] he wrote in an article that provoked the National

Landowner's Association to complain to Funk of its strident tone. He sent a telegram to Hitler on 4th April, calling for a radical turn to government policies. Looking back after the war, Darré seemed the logical choice for minister[32] but at the time he had to force his way through to government level.

One means was the Prussian parliament, which provided Darré's first victory against Hugenberg.[33] In April 1933, Darré and Backe met him to try to win his support for the proposed new marketing laws, and to discuss the new Prussian Inheritance Law. The latter was to provide a 'dry run' for a national Inheritance Law. Hugenberg refused to give way on either proposition. He disapproved of institutionalising controlled prices, and agreed with von Rohr on the potential problems of the Inheritance Law, which would immobilise land transactions. Goering, Prime Minister of Prussia, was approached instead.[34] Hans Kerrl, the Prussian Minister of Justice, 'seized the initiative' on 15th May, 1933, after a conference with Willikens, and pushed the Prussian Inheritance Law through the Prussian Parliament against the opposition both of Hugenberg and the Prussian Minister of Agriculture.[35]

This was a triumph for both Darré and Backe, who had jointly drafted the law, and Darré later wrote generously of Hans Kerrl's role in 'performing a historical service'.[36] On 4th May, 1933, Darré was elected the chairman of the council of all agricultural organisations, the post filled by Brandes when Darré had been job hunting in 1928. He promised Backe that he would be State Secretary as soon as Darré had control of the Ministry for Agriculture. Backe was still his 'closest co-worker'. He wrote that 'Darré was wonderful', although he found many of Darré's personal staff inadequate, there either because of their name and status, or because of past services to the Party. Darré's ability to make rapid decisions and think ahead was impressive, and Backe felt that at last 'he had worked at something important' with the Prussian Inheritance Law.[37] In fact, the Prussian Inheritance Law had to some extent legalised existing north German primogeniture customs. The real problems lay in the future, when the law was extended to areas of Germany that had known the Code Napoleon for nearly 130 years, and were used to multiple sub-division of the land among the heirs.

Between May and June 1933, Hitler was pressed by Schacht, von Papen and Hindenburg to retain Hugenberg. 'For Hitler, my opposition to Hugenberg is a problem: I have the peasants, while Hugenberg has the position of power that he built up,' wrote Darré. Hindenburg invited Darré to visit him in order to ask him not to 'destroy the great estates'. The visit went well socially, as the two men found a common enthusiasm for the Hanoverian horse; but nothing was agreed on the vexed issue of land tenure.[38]

However, as mentioned earlier, Hugenberg's attempt to outdo the Nazis in his London speech alarmed his fellow conservatives, and in June 1933 he resigned from the Government, giving Darré and the AD a free hand. On 14th July 1933, the *Reichsgesetz über die Neubildung Deutschen Bauerntums*, the Law for the New Formation of a German Peasantry, was passed, giving the AD in its new guise of Agricultural Policy Office a free hand to organise peasant settlement. However, it appeared that a bargain had been struck behind Darré's back. Despite his apparently powerful position as head of the Ministry of Agriculture and national peasant leader, the Law for the New Formation of the German Peasantry limited settlement to land which was available for purchase on the open market, plus land owned by the state, in need of reclamation. Nor were the necessary funds forthcoming from the Finance Ministry.[39]

The National Inheritance Law increased pressure on the settlement department. The younger sons of those farmers now subject to the rules of primogeniture had been promised new farms and smallholdings as compensation for their loss of prospects. Nonetheless, the AD refused to be deterred by the stumbling block over funds. They hoped that the shortage of money would be only temporary, and during August and September Backe and Willikens worked steadily on the draft law. 'As peasant leader Darré is splendid, he has instinctive certainty', Backe wrote in August. Even a year later, his wife could note in her diary, 'Herbert says Darré's work will be seen as the great success of National Socialism ... the RNS staff are realism itself ... without Darré's vision and large-scale political aims, nothing could have been achieved.'[40]

After enthusiastic discussion, Goslar was chosen as the new peasant capital. Goslar was an early medieval town in the foothills of the Harz mountains, the capital of the Holy Roman Empire in the tenth century. Its history as a mining, farming, independent city state conformed to Darré's ideal of the republican freeman, and suited his admiration of the Renaissance city states. He knew Goslar from his sister's schooldays, and his mother was buried there. Furthermore, the countryside around the little town was unspoilt; gently rolling hills where sheep grazed, while to the south stretched the Harz mountains. He decided on Goslar as a suitable centre for the rebirth of peasant Germany after examining photographs of this serene landscape. Peasant festivals were held on the slopes of the Bückeberg, rallies attended by up to half a million farmers, and often addressed by Hitler. The National Food Estate's meetings were held in the hall beneath an old hotel, once the town hall. It still has its medieval paintings on the beams. The dream was to make Goslar the centre of a new peasants' international; a green union of the northern European peoples. Here he made speeches condemning the 'Führer' principle and attacking imperial expansion. Visitors flocked to him. Organic farming enthusiasts from England now welcomed Darré's plans, and admired the hereditary tenure legislation. Representatives from Norwegian and Danish peasant movements joined the conferences on 'Blood and Soil'. The real heart of the muddle of agricultural ministry, National Food Estate, NSDAP Agricultural Office, Peasant Leader's Office and personal Minister's office lay in the National Peasant Council's Keeper of the Seal, and this too centred on Goslar. Here, the personal files of all members, who included the *Reichsleitung*, Gauleiters and all *Landesbauernführer*, were kept, together with correspondence with the far-flung, decentralised National Food Estate offices. The first holder of the post was an old Halle friend of Darré's, Richard Arauner, who was killed in a car crash in 1937. He was succeeded by a Gauleiter Eggeling.[41]

But as early as 1933, the first hint that Darré's large-scale aims might become out of step with those around him had already appeared. 'Economically, he is inclined to erect ambitious intellectual structures— getting things basically right, because he can see ahead, but the flight of his thoughts is too quick. He sees at once aims which could only be reached in a century or in decades'. Backe's assessment was right. Darré's very ability to stick to his radical views, which had been such an asset in the *Kampfzeit*, was now becoming a liability. He refused to admit that the revolutionary time was over, and that for his companions the time for consolidating power and power bases had come. When Darré and Backe visited Schmitt at the Ministry of Economics in early September 1933, to discuss the proposed Debt Relief Law, Backe found Darré's unexpected remarks about trade policy and tariffs an embarrassment. While he regarded Schmitt and Posse as politically 'unsound', he sided with them when Darré expounded his radical proposals to de-industrialise Germany, to leave the cities to decay, and to concentrate resources on the land.[42]

However, Darré still supported Backe as von Rohr's successor. Von Rohr opposed the Inheritance Law in a Cabinet discussion of October 12th, 1933, at which Darré's survival as Minister came into question. It was in part due to Backe's persuasive lobbying of Schacht and other opponents that the Inheritance Law went through, and Darré dedicated the first copy of the National Law Journal containing the EHG to 'my dear Backe, to whom the birth of this law owes so much'. Both men considered this law, and the National Food Estate Law, which erected a marketing board to supervise and purchase all primary and secondary agricultural produce, as the 'liquidation of Hardenberg' (Darré), the 'complete demolition of the last 150 years' (Backe), and so on. 'The Inheritance Law is the deathblow to Reaction, and all that is implied by the large landowner [*Grossgrundbesitz*]'.[43]

To some extent, this was wishful thinking. What Darré had done was to give small to medium sized farms (originally seven to 75 hectares) the same legal protection against foreclosure and sale as the large entailed estates had. This did not make their competitive position stronger. Darré had not even begun to strike the deathblow at the Junkers which he sought, and which the Junkers feared. Six years later, he managed to pass the Entailed Estates Law (*Fidei-Kommissgesetz*) which dissolved the remaining 1400 or so entailed estates. This act, by 'ending the feudal millenium', introduced the capitalist one, rather than the allegedly idyllic pre *Bauernlegung* era. By removing all restrictions on use, pension obligations, life interests, and other barriers to land mobility and alienability, fully capitalist relationships were introduced to entailed land-holdings. Junker power did weaken, but this was due to the complete shift in power structures under the Third Reich. The new armed force, the SS, the increase in meritocratic ideas, the attempt to open up the civil service, and the increase in government posts, all demanded a more open recruitment policy. The middle management class, the foremen, the scientists, the NCO and Major levels in the army, these produced Hitler's most devoted followers in the 1930s. Agricultural legislation by itself played little part, despite Darré's sincere intentions.[44]

In any case, entailed estates were a symbol rather than a real force: they formed a numerically tiny minority of estates over 500 hectares, although a higher proportion were in noble hands. Here Darré was a victim of the current presumptions of his time about the backward, feudal nature of the Junkers, while at the same time attributing Junker power to the Hardenburg reforms of 1811–16, a myth that persists to this day. The result was a strange kind of shadow play. Although Darré no longer represented a real threat to large landowners, once his land division plans were passed over, he was still feared. He attacked the Junkers fiercely in 1934 at Starkow, to the enthusiastic applause of

30,000 peasants. He complained of an 'agricultural plutocracy' in a speech to the German Agricultural Society in 1933. In January 1935, a new Settlement Act was passed, which increased the pre-emptive powers of government land purchasing agencies, although not increasing their funds. Small wonder, then that Darré was feared by the conservatives, and admired by his fellow radicals, despite the lack of real result.[45]

Once the 1933 legislation was passed—in the case of the National Food Estate, with remarkable ease, considering the still somewhat Conservative cabinet—new problems emerged. Backe's enthusiasm began to wane. The creation of the marketing system had been satisfying: the job of enforcing it was more difficult.

The first obstacle was the heavy debt burden carried by German farms. Backe had planned a *Umschuldungsgesetz*, which would have divided up agriculture's debt burden equally between all the *Erbhöfe*, but this was too much for a partly Conservative cabinet to swallow, and the law was thrown out on 14th October, 1933. Backe was mystified by the hostility to the draft law. Farms in the east were three times more in debt than their equivalents in the west. Eastern farms which had invested heavily in machinery and increased productivity were the worst hit. He failed to understand why farmers in the west should object to 'sharing the burden of their brothers in the east', and taking over their debts at a low rate of interest. When the law was dropped, and a plan to fix interest rates at two per cent adopted, Backe observed critically that this showed that Hitler had dropped his corporate ideals, and was beginning to think from a nation–state viewpoint.[46]

Other difficulties arose. The staff needed to control production levels and distribution was a huge expense which was funded by the farmers themselves. Furthermore, Germany had begun 1933 with low food reserves, and although the 1933 harvest was good, that of 1934 was poor. Darré's reaction was to ask Schacht for foreign currency to buy fodder for livestock, to enable the 'fats gap' to be closed (Germany produced about 50% of her fats requirement). He was to wage a three-year battle of speeches, letters and memoranda against Schacht, who was determined to avoid inflationary measures.[47] He attacked Schacht for economic liberalism at the 1934 *Bauerntag*. He told Hitler that Schacht was to blame for problems with foreign trade. However, Hitler received these tirades with passive inaction, which became active resistance in mid-1934. Goering and other party leaders began to criticise the marketing corporation for lack of co-operation, a hostile attitude, and inefficiency. Darré's request for a further price increase for livestock producers was refused by Hitler. The request involved a loss of face for Darré, since he had fended off a previous concerted attack by a group of Gauleiters by telling Hitler that cereal prices (and hence bread prices) should *not* be raised, knowing that Hitler supported a policy of cheap food for the workers.

Darré and Backe had requested this meeting with Hitler, which took place on 5th July, 1934.[48] Darré was concerned with prices for producers, while the attacks on him concentrated on consumer prices. At this stage, one of life's basic problems had emerged in all its stark reality; that you cannot have a policy of dear food and cheap food at the same time. During the change-over to distribution controlled by the RNS, wholesale-retail margins shot up temporarily, as reports up to 1936 show. It has been argued that official figures do not reflect the true rise in retail food prices, but if agricultural producer price indexes are broken down, it can be seen that wheat prices fell to 85 in the year 1933-4, from 91 in 1932-3 (taking 1928 as the base year = 100), while livestock prices rose from 53 in 1932-3 to 59 to 1933-4. Darré felt that this was not enough to help the small farmer/livestock producer. In 1934-5, the livestock index rose to 72, and the following year, to 90. Another reason for Darré to favour lower grain prices was that Germany was virtually self-sufficient in grain, and increasing grain production was not a major policy aim. Of course, grain growers were also associated with the hated east Elbian farmers and their subsidies of the late Weimar period, so that wheat price movements between 1933 and 1936 were untypical of other agricultural prices.[49]

Hitler displayed distinct signs of boredom at the meeting between Darré, Backe, himself and the Gauleiters. He took up a newspaper and began to read. He asked Darré jestingly if he had 'mystical-romantic' feelings about the bread price, and when Darré replied that he wanted it lowered, he said, 'Good, that is my opinion also,' and closed the meeting. This left matters hanging as far as the balance of power between the National Food Estate and the Gauleiters was concerned.[50]

Darré then sent Backe, in August 1934, to discuss food prices with Dr Ley, head of the German Labour Front. The Trustees of Labour had a two-fold conflict with the National Food Estate. First, there was a clash of jurisdiction regarding agricultural labourers, and second, the trustees wanted the cheapest possible food for their labourers. But Backe drowned out Ley's complaints with a contemptuous lecture on the workings of supply and demand.[51] Also in August 1934, Goering launched an attack on the Ministry of Agriculture. He collated 30 pages of complaints from the Gauleiters of thirty-one districts, accusing the agricultural officials of independence, interference with Party matters, and disloyalty. Goebbels, more seriously, had accused the RNS in January 1934 of not supporting the *Winterhilfe* programme, a charity which provided food and clothing for the unemployed; it was run by Party workers, and played an important role in maintaining the Party's image as friend of the working class. Darré produced a 47-page rebuttal, sought a meeting with Goebbels to discuss the charges, and then squabbled about the administrative 'hat' Goebbels was wearing. Goebbels walked out.[52]

Hitler installed Goerdeler, ex-mayor of Danzig, and later participant in the bomb plot of 1944 against Hitler, to be Price Commissioner until July 1935. Goerdeler recommended lower agricultural prices, a return to the free market, and the elimination of the inefficient farmer. This was consonant with what is known of Hitler's own economic views, but was blamed on Darré's inability to interest Hitler sufficiently in agricultural problems.[53] Goerdeler's plan was an attack on the RNS ethos, since the agricultural advisers were meant to substitute education and advice for the self-correcting mechanisms of economic competition. Both Backe and Darré smarted under Goerdeler's attacks. In an attempt to find allies, Darré was even driven to compromise with Ley, in 1935, in the so-called 'Bückeberger Agreement'. According to this arrangement, the National Food Estate joined the German Labour Front, and replaced the Labour Front's agricultural labour section. They agreed to contribute funds in exchange for sharing some of their welfare arrangements, such as supplementary unemployment insurance, free holidays, and so on. This move was unpopular with farmers, who were not anxious for their agricultural labourers to join an external and powerful body. Goerdeler fell on this sign of weakness, and drafted a law whose effect would have been to dissolve the National Food Estate, under the guise of simplifying the corporation's structure.[54]

Backe was taken by surprise by Darré's vacillation over Goerdeler's move. He waited for Darré to go to Hitler and demand Goerdeler's dismissal, and when he did not, decided that Darré was 'a flop: internally weak and insecure, and lacking in courage'. Backe by now had moved away from Darré's long-term and supra-political aim of a peasant-based Germany, which would lead to the 'racial renewal' of the German people. The 'Battle for Production' had been launched in November 1934, and this led him back to his main interests of the early 1920s, the belief that Germany's agricultural problems were inextricably bound up in the world-wide economic system. He wrote to 1936 that he was searching for 'an economic principle that would embrace the whole of the economy, not just agriculture', and tried to define his ideas by studying economists, such as Sombart. His concern was about Germany's lack of self-sufficiency in raw materials and some foods. In order to create a new economic system, Germany had to be able to survive without dependence on multi-lateral foreign trade. This was not so much because of boycotts against Germany, the loss of colonies, etc., but because Backe saw this as the 'new' system, a 'new' idea, into which Germany, first of all nations, had been forced by post-war circumstances. To define the 'new idea' precisely proved more difficult, but as it was seen by Backe and his circle a matter more of spirit and will than of pedantic definitions, the imprecision did not matter.[55]

At the highest party levels, Backe's success in the fight for higher productivity was noted, and his aggressive nationalism approved. Backe was sociable, unlike Darré: a good companion with whom to drink and talk for an evening. He was loyal, intelligent, and above personal ambition. In 1936, he was appointed agricultural representative to Goering's Council of the Four Year Plan.

The genesis of this appointment was ironical. Darré had asked the Ministry of Economics in March 1936 to set up an enquiry into foreign currency, raw materials and food, as part of his battle against Schacht and the share of foreign reserves allocated to agricultural imports. Goering was put in charge of the investigation, with Schacht a last-minute substitute for Blomberg. This 'fixed' the committee, while Schacht wrote a memorandum on foreign exchange on 23rd April 1935. Early in May, Hitler ordered Goering to form a new committee on raw materials to examine national autarky. Both Darré and Backe attended the first meeting of this committee, at which 19 Ministers were present.

In June, Parchmann, an official in the Forestry Office, and a keen admirer of Backe's, recommended him to Goering as a gifted economist. Backe now represented the agricultural sector on Goering's committee on the organisation of raw materials and currency, which met on 6th July, while on the 15th, he was formally nominated to the Council. The Second Four Year Plan Council was convened on 18th October 1936, and Backe was appointed agricultural representative. This marked a shift of power from Darré to Backe. For other ministries, figure-heads were co-opted to liaise between the council and the ministers, but by placing the gifted and articulate Backe in this role, Darré had effectively been demoted. 'The personal union between Backe and the Food and Agriculture Committee means that (there) will always be a lever against me', wrote Darré, who pushed first a National Food Estate nonentity, then himself for the post. He added in his diary, in his first criticism of Backe, 'Now Goering will understand what I have to put up with. Backe's Russian nature ... talk, talk all night, never action'.[56]

Were Darré's fears for his position alone? One problem with tracing these power plays and strategies is that it is easy to assume that personal position alone was at stake. But here Darré was aware that he was virtually the sole 'bearer' of the peasant ideal. If he went, there would be nobody to take his place. However it was also obvious that if he stayed, he would be a lone voice. The day of drastic social re-shaping, of revolutionary fervour, was gone, and ideas of ignoring the worker interest in favour of the peasantry were not popular, especially with Hitler. Well aware of Gauleiter hostility to the RNS, his especial creation, Darré feared that Backe might be prepared to renounce its relative independence and corporate structure. Certainly, the Four Year Plan, with its centralised planning and war economy, was a death blow for peasant socialism.[57] Backe realised that a modernised and highly organised economy was the natural development for the 'new Germany'. He was attracted by the prospect of a German empire, while Darré recognised the 'preference for an empire over internal reform' as a defeat for his beliefs. However, he had lost his touch for political campaigns. Not realising this, he kept hitting out at his foes, often using irrelevant quibbles about competence that other ministers found tiresome. He was perhaps vain enough to want to cling to office, despite the expected disaster, but not sufficiently vain to lay aside his beliefs for the sake of ambition.

The constant arguments and frustration affected Backe's health, already weakened by his hardships in youth, and he spent some time in a sanatorium for heart cases. Goering became concerned at the extent to which his usefulness was affected by the disagreements within the ministry, and told Backe to return to Berlin to discuss the matter. Backe's main complaints at the ensuing meeting were that Darré could not control his staff, who were pretty useless anyway, and that Darré had no understanding of economic questions. He forcefully criticised Darré's indecision and vacillation, his lack of judgment, and his tetchiness. Goering was obviously tempted by the idea of getting rid of Darré. He mused that it would be a mistake to assume that the National Food Estate had the whole-hearted support of the peasantry—even Himmler, Darré's great friend, would agree to that. However, he finally decided, regretfully, that it would be impossible to sack a Nazi minister without damaging the reputation of the other Nazi ministers.[58]

While Darré sat at home nursing an Achilles tendon strained in a 400 metres race, Backe visited Himmler at Tegernsee, in September 1936. He found him 'impressive, but humourless and tense'. Himmler repeated Goering's point about the need for unity as far as the public were concerned. A lack of unity had been the rock on which all such movements had foundered in the past, and he saw it as his special task to maintain unity whatever the cost. After Himmler stressed Darré's past services to the party, the fact that 'he had been absolutely necessary for the movement', Backe cooled down his complaints. He now directed his attack to Darré's subordinates, winning Himmler's favour in this by accusing them of wanting to raise agricultural prices—a move Hitler opposed ('Himmler especially impressed'). Himmler ended the session by promising to have a word with Darré, and warning Backe against Moritz, one of Darré's staff members, because he was suspected of being Jewish.[59]

Darré heard about these visits, and responded vigorously, but soon found that it was one thing to lash out in 1931, attacking the *Landbund*, attacking *Die Tat*, and the Stahlhelm because of its Masonic connections, and quite another to go for the top Nazi leaders in 1936. He could not even protect his own men. One *Kreisbauernführer* was dismissed and imprisoned for six months in 1935 for repeating an allegation heard at a Goslar Peasant Festival that Blomberg and most of the senior army officers were Freemasons.[60]

For Darré, it was a reversal to the 1926 days when he sent secret reports on corruption and treachery to unconcerned, uninterested committee members. His judgment had gone. For example, in response to a purely formal note of condolence from Goering (Darré was bedridden for several months in 1936) he replied with inappropriate condescension, congratulating Goering on his progress with the Four Year Plan, assuring him of his support, and promising him the benefit of his advice whenever necessary. He then set down his considered judgment of Backe's faults. That he should do so to Goering, a Backe supporter and keen opponent of himself, shows a lack of common sense, and suggests a considerable isolation from ordinary human relationships. Unlike Backe's letters, Darré's outpourings were smoothly written and articulate, drafted, redrafted and polished, but their contents were painfully misplaced and misdirected. Darré may have realised this to some extent: there is an air of justifying himself before posterity, regardless of the effect on his correspondent.[61]

In his three-page attack on Backe, he praised his 'academic intelligence', but criticised him for a lack of political awareness, this supposedly a result of Backe's Russian upbringing. The core of the current argument was that Goering had asked Backe to collect some statistical material from the RNS, and Darré, unable to produce it, had flown into a rage. He had a genuine defence; collation of statistics had been handed over to the National Statistics Office in order to avoid duplication of effort. Peasants had complained that they were financing a purely governmental activity via 'their' corporation. Each point that Darré made in his own defence was logical, and, taken in isolation, even relevant, but the absurdity of writing a droning, 16-page letter to a busy Minister, one who was not noted for his tendency to study documents closely, and one who would certainly not support the writer on the matter concerned, displays, more clearly than all Darré's letters at this time, the reasons for his growing isolation.[62]

He drafted his letters shut up alone in an office, insisting on the best standard of typing, just as he insisted that his SS bodyguard should always be the prescribed military distance behind him as he went to his office. But contemporary photos and newsreel film show a soft-shelled, uncertain exterior that could not have fooled his contemporaries.

Backe's visit to Himmler at Tegernsee had occurred at a time when Hitler was drafting the drive for economic self-sufficiency which was later to become the Four Year Plan directive of 18th October 1936. While Backe's appointment divided the Ministry of Agriculture staff, the memorandum of 4th September 1936 united them in a belief that Hitler was at last relinquishing the shreds of the market economy that had persisted since 1933. Darré exulted that Hitler had 'made a thorough-going attack on economic liberalism which left Schacht perplexed and helpless', while Backe heard from Paul Körner of the Ministry of Economics on 7th September, announcing that Goering had received 'new guidelines for all our work' from Obersalzburg. Both men saw Hitler's action as expressing a commitment to the *gebundene wirschaft*, and the imminent break-up of the agrarian radicals was halted. Backe felt that Hitler's speech of 4th September—'We are not concerned with a bit of butter here, or more eggs there, our first duty is to ensure that the broad masses of our people can work and serve, and protect them against the horrible suffering of unemployment'—was a 'wonderful vindication of three years' hard work'. Even when addressing the Wehrmacht, Hitler took the opportunity to emphasise the Socialist element, and Darré extracted the juicy bits for *Landpost*:

No nationalism can really exist right now which is not determined in a Socialist way by the *Volk* community. And no one is a true National Socialist for whom the emphasis fails to lie on the word 'Socialist'. Here, in this concept, is what lends impetus to our era.[63]

Hitler's apparent rejection of economic liberalism temporarily consoled Darré for Backe's rise to power, and both men had high expectations of the proposed Four Year Plan. Darré hopefully compared Goering to Friedrich List. The weakening world trade system, the growing threat of sanctions against Germany, seemed to make autarky inevitable. 'He whom the world rejects must reject in return—because in the world economy, goods can only be exchanged against goods', wrote Backe. However, this simple law held good within Germany, too, and Goering's management of the Four Year Plan failed to be the expected bonanza for the agricultural sector. The council was not the expert committee of unbiased technocrats which was the core of National Socialist economic ideals. Backe complained that because of Goering's 'optimistic dictatorialness', many important decisions made on paper round the council table were never put into effect. The 'same old party hacks' were appointed, instead of the council of experts. Furthermore, Goering's views of agricultural priorities were affected by an apparently trivial matter: his devotion to the chase. Constant complaints from farmers whose lands had been damaged by hunting landed on Goering's desk. This infuriated Goering, and gave him no reason to share the Ministry of Agriculture's commitment to a successful class of small farmers. Goering, who had taken charge of the Forestry Office, did create nature reserves and landscape protection areas in Germany, many of which still exist. But his lack of sympathy towards the agricultural sector was affected also by his dislike of Darré, whom he considered an impractical mystic, Darré's opposition to the large landowners, of whom Goering was one, and, perhaps more importantly, by his realisation that Hitler had lost interest in the peasantry.[64]

Backe's own ideas for improving Germany's food supply, including land reclamation, the introduction of *Volk* fish farms, increased mechanisation, artificial fertiliser use, and so on, seldom received more than a bored response around Goering's council table. It was impossible, he complained, to interest Hitler in silo building, or the need for modernised dairy production. At one showpiece event, the International Dairy Conference of August 1937, attended by academic delegates from all over the world, Goering made a speech that left a distasteful impression on the autarkic agrarian radicals:

> No country can withdraw today from the world economic system. No country can ever say again, we decline the world economy, and are going to live and produce for ourselves alone.[65]

Nonetheless, Backe could share in the vision of a strong new Germany, and had enough in common with the top Nazi leadership to be able to co-operate with them in war preparations. Darré could still produce sensible, if long, policy documents on productivity and other practical issues, but his known belief in de-industrialisation and massive social change made it inevitable that he would be passed over by Hitler and Goering, even if he had not begun to undergo a serious deterioration in personality at this time. Backe, though, was admired and while nobody was prepared to listen seriously to his proposals, there was a general gut conviction that Backe, being a practical sort of man, could definitely produce food when it became necessary. How, precisely, would be his problem. Backe responded emotionally to this expectation: dutifulness and the need to serve, to belong, drove him on. 'I have to force myself to work harder and harder, because I feel the necessity of achievement,' he wrote revealingly. 'But Darré lacks this discipline. I still have something to achieve, and I *must* live towards it ... I have to go this way, even if the decay and ruin of my former travelling companion Darré means that he can no longer walk with me.[66]

But Darré had not yet chosen the path to total retreat. On his return from a long illness in January 1937, he appealed for loyalty. He was aware of the problems of a division of authority, which had been brought about by Backe's role in the Four Year Plan council. The fragile network of authority which strung together the RNS, the Ministry of Agriculture and the peasant leader depended on the fact that one man headed all three. Darré's first act back in the office was to circulate a memorandum calling for unity and support to Backe and his colleagues. The support did not materialise, not because Darré's staff dislike him, but because they were genuinely perplexed as to whom to obey. Several meetings called by Darré in February 1937 broke up in disagreement, sometimes angrily dismissed by him. For example, one meeting was convened to set up a working party to discuss the increases in productivity demanded by the plan. The group decided that the marketing corporation was the most suitable instrument with its established local network of farming experts. But Backe then announced that while he could accept Darré as chief in his (Backe's) capacity as State Secretary, he could not do so as a member of the General Council, and that this should be clarified as a preliminary to any action which might be taken. Darré, with some justice, remarked that one could not have two ministers running the same ministry, and dismissed the meeting.[67]

Darré was now retreating into the National Food Estate, which he saw as a bulwark against economic liberalism, that enemy to Blood and Soil and the peasant ethic. Any attack on his own authority was seen as an attack on its crucial role. He had the support of the lower-level staff here, but proposals for improving its efficiency were ignored by the four State Secretaries. It is an interesting comment on the nature of the totalitarian dictatorship of the Third Reich that a man with so many offices, titles and apparent power was as unable to enforce cooperation from his civil service equivalents as would be a minister in a democratic government. Darré argued that agriculture faced problems that were insoluble because of extraneous factors. Two bad harvests running had caused concern to the army. Germany's balance of trade was worsening, and rearmament was causing a shortage of agricultural labour. No marketing system in the world could cope with a shortage of land, men, machines *and* of produce. Nor could these evils he remedied without structural change at the top; the policies of Schacht and Goering would have to be altered.[68]

Given his basic premises, this was a justified analysis. But Backe considered the demand to be wholly unrealistic. Darré responded that to attempt reform within an unsuitable political structure was unrealistic. The result was that both men began first to suspect, then to despise the other. While attacks from other party organs could still produce unity (for example, Goering's proposal for a grain monopoly in July 1937 found a united defence), more fundamental divisions between Darré and the Nazi Party were beginning to surface. Backe tended to confine his criticisms to Darré's character, not understanding the psychology of a man who could feel defeated from the outset, yet be unable to retire. He blamed Darré for not approaching Hitler more often when the RNS came under fire, attributing this diffidence to moral cowardice. He decided that Darré was probably afraid of being 'caught out' in factual questions, and preferred an isolation which Backe believed was damaging the National Food Estate's already precarious position.[69]

It is true that Darré liked to rely on Backe to produce telling statistics at short notice. On a visit to Italy before the war, Mussolini asked both men what proportion of Germany's population were peasants. It was Backe who returned the figure, one, after all, crucial to all Darré's policies. Mussolini, incidentally, was horrified at the answer. How could a nation produce enough food for itself if only 25% of its nation were peasants, he asked rhetorically.[70] But Darré's hesitation about approaching Hitler over the heads of other *Reichsleiter* was because he had never enjoyed Hitler's close confidence and trust since the first meeting with him at Saaleck. At root, Darré was a one-issue man, and he liked to use Backe as his front man, because he knew how committed Backe was ideologically and emotionally, to National Socialism, which he perceived as Hitler's attempt to implement the 'New Idea' which he thought had germinated in the apparent collapse of western capitalism after the First World War. This entailed a belief that the survival of the National Socialist power structure was all-important, while Darré retained sufficient objectivity to criticise the path Germany was following. He had, after all, made a conscious decision to use the NSDAP for his own beliefs back in 1930, and behind his depression and lethargy of 1936-7, the flood of quarrelsome memoranda, the endless drafts for letters designed to rescue inconspicuous personal points of honour, lay his growing realisation that he had made a terrible mistake.

His very successes of 1931-3 proved ruinous to his later political manoeuvres. They gave him the impression that forthright aggression would always triumph. The Christian *Landbund* leadership, the Raiffeisen co-operatives, and all the other bodies which had fallen like ninepins under the combination of his infiltration programme and peasant discontent, were a different problem from the Nazi leadership. Men who had been demoralised by war-time defeats and post-war events, and who lived in the expectation of failure, could be bluffed into defeat; Himmler, Goering and Hitler, with a record of success and an ethic of complete ruthlessness, could not. Darré found that by 1937, with the peasant vote no longer needed, he had lost his power base. His wishes were circumvented or simply ignored.

Yet in many ways, this man, regarded now as a complete mystic, a fantasist, a romantic and a dreamer, was fundamentally more clear-sighted than the Nazis. He had never fitted in easily with institutions and structures, and was unable to regard them as important. Probably he underrated the value of the cohesion, continuity and security that can arise from such bodies, but then that would have been a conservative position, and Darré was not a conservative. He had seen political frameworks undergo revolution, not just in 1933 but in 1918–1919. They were not sacred. Radical change had occurred through what seemed to him the general recognition of pressing necessity, and he saw no reason why it should not happen again, when such a change was clearly necessary to produce the desired effect. Governments were merely man-made, and subject to variation at human will; the laws of agricultural productivity, of historical development, were not.

An example of this attitude is Darré's apparently irrelevant canvassing of the need for organic farming during the 1930s and 1940s. Backe was irritated by Darré's insistence on a long-term policy to improve the humus content of Germany's soil ('Darré muttering about organic farming'), and continued to call for nitrogenous fertiliser to be made available for agriculture. Supply failed to improve. Nitrogen was actually a net export during the war, and productivity dropped after a war-time peak of 1943. If his arguments had been heeded, a long-term programme of soil improvement could have been put into effect in the late 1930s, and helped agricultural production during the straitened war years.[71]

Another ideological divide became apparent in the emergence of Germany's territorial ambitions. Although it might seem obvious in hindsight that National Socialist Germany would be forced to seek foreign conquests—to obtain raw materials, to fulfil party promises, to relieve inflationary pressures—Darré seems to have believed that he could influence the party away from its militarist as well as its urban elements, perhaps because his point of contact in early Nazi days had been the peasant-oriented Thuringian branch. Darré was not to support the invasion of Russia, and he was equally shocked by the Ribbentrop-Molotov Pact, which he thought gave too much strategic land and railway access to Russia, and displayed over-Machiavellian cunning on Hitler's part. Darré protested to Hitler in 1937 when the latter began to talk of his Russian ambitions: he would confine settlement projects to German territory, together with settling East Prussia and the Baltic area more heavily with German farmers. 'The *Ostsee* is our imperium', he argued, and was an early supporter of the return of the ethnic Germans from Russia for German settlement.[72]

His support for the Four Year Plan as a 'blow to liberal economic thinking' faltered when the link between autarky and preparations for war became clear. War meant the 'non-agricultural economy'. In 1936, the outcome of what was seen as a choice between full-scale 'foreign adventurism' and the controlled economy seemed uncertain, and Darré speculated as to whether Goering would go one way or the other. But when policy appeared more defined, Darré wrote that he would 'lay aside my earlier source of strength, the belief that I was working for a German Revolution, as an idealistic Utopia', and in April 1939, 'As a result of our foreign policy, there is such a brutal, heavy industrial economic imperialism abroad, which makes one anxious for Blood and Soil ideals—Germany a colossus with feet of clay ... the Protectorate will be a failure without some common tie of blood [*Blutgedanke*]'. These are phrases which spell out clearly the inherent opposition of national to imperial ideas.[73]

On the other hand, although Darré had fairly consistently conceptualised his ideal state, and could see where Nazi Germany's developments opposed it, he had failed to think through the problems of achieving that ideal state while not destroying the ideals in the process. The problems of revolutionary aims versus revolutionary methods has perturbed political revolutionaries from Plato to Lenin, and is too large a problem to analyse here. Nonetheless, it is fair to ask exactly where Darré thought the 'Baltic' ended, and what kind of struggle he envisaged in order to gain control of it. How did he imagine the *Ostsee* was going to be made available to German peasants without creating the necessary tools for achieving the task? How was war to be limited to acquisition of the Baltic? Not just factories, tractors, nitrogen, tanks, aircraft, but the war machine capable not merely of conquest but of the resettlement of an already occupied territory, together with the sustenance and subsequent development of the settlers—all were to be provided from a home base poor in raw materials. It should have been clear that the very process of conquest would have corrupted the conquerors; that the men who were to carve out a living from the land for the sake of Blood and Soil would have found it hard to resist the spoils of war. The very faults that Darré had diagnosed in German society, the unco-operative, vulpine competitiveness, the tendency to retreat to a deliberately unintuitive bureaucracy, the loss of faith and sense of defeat among German farmers, could only be enhanced by his prescriptions.

There were three areas where Darré *was* prepared to 'go to Hitler'— if he could force him to receive him—and these were peasant settlement, the shortage of agricultural labourers, and organic farming. He was given short shrift on all three.[74] Soon after the first agricultural legislation of 1933, Darré called for a voluntary code among farmers, through which labourers would be provided with a smallholding by their employers, in exchange for one day's work a week. The labourer's holding was supposed to be large enough for self-sufficiency. Darré wrote at this time that the 'new modernity' lay in farming units which were designed to produce the maximum self-sufficiency, because the day of mechanised cash crop farming was over. In 1937, he called for a halt to industrial expansion, and in July, visited Hitler to discuss the 'catastrophic shortage of farm labour'. He suggested either importing ethnic Germans to work as agricultural labourers, or slowing down industrial expansion. He wanted legislation to stop the *Landflucht*, the flight from the land, but Backe and Himmler, who were present at the meeting with Hitler, were contemptuous. Darré continued to write memoranda attacking the importation of Polish labourers. The thought of bringing in Poles to fill the place of German peasants who had gone to the factories was intolerable to him. It was a bitter irony that a party which he had joined to further the peasant cause should preside over its apparent destruction. One historian calls Darré's diagnosis of this issue 'as realistically based as its conclusions were unrealisable; to cut arms production drastically, to remove inflationary overheating from the economy'. This description could be applied to many of Darré's campaigns: realistically based, but unrealisable.[75]

Hans Kehrl records a speech given by Darré in April 1937 to an educational conference. He bemused an originally sympathetic audience by discussing 'philosophical theory' in an 'unclear and wordy fashion', losing the interest and sympathy of his listeners. Hitler was reported to have complained about Darré's 1937 speech at Goslar. 'The peasants certainly aren't interested in listening to all that peasant philosophy stuff'. Darré had obviously lost his political instinct. His preoccupation with theory reflected his knowledge of the failure of his mission, his realisation that his convictions were no longer shared by others, and his need to persuade them.[76]

For Backe, public alignment with Darré was now an embarrassment, although he was well aware that agricultural productivity increase was faltering in the late 1930s. He did not take part in Darré's fight for the abolition of the entailed estate in 1938. The abolition of the entail was another paper victory for Darré, and probably a gesture made to the radical agricultural sector by Hitler in a year in which Party unity was particularly needed. Darré saw his law as a threat to the Junkers, and so did they. They flooded him with social invitations, and for two months, while Darré drafted the law, he was wined and dined by various members of the nobility. He even received invitations to meet Goering at the Karinhall. He agreed with Goering to provide a protection clause for private forests, to prevent their division and sale. Goering then sided with Darré at a meeting with the Justice Ministry, and against Gürtner, agreeing that 'dynasties should not be put into cold storage.'[77]

Hitler signed the law of 6th July 1938, but it is doubtful if he would have done so if he had expected the *Fidei-Kommissgesetz* to create havoc among the large landowners. After all, the entail had prevented the full utilisation of land. Entails meant pensions, provisions for care of dependants, obligations to families and employees. The new law's provisions for compensation for the loss of legal interests in the property were designed to ensure that the estate remained a viable farming unit. If the holder of a life interest could not be compensated without stripping the estate of some resource necessary to its productivity, then the holder went without compensation. This was the extent to which the law envisaged the need to maintain the estates as viable production units. As argued earlier, the effect of this law was to remove the vestiges of neo-feudal community obligations, and institute maximum flexibility of tenure and alienability of land. If the law had been a real blow to large farmers, it is doubtful whether Hitler and Goering would have agreed to it. In 1933 the large estates had been poor, ripe for bankruptcy. An attack on the entail then would have meant forced sales, sub-division and settlement. By 1939, most large farms were more profitable, and land prices had risen. Despite the fact that Darré regarded this law as his last achievement, a final blow against his enemies the capitalist, anti Blood and Soil landowners, all it did (or would have done if the war had not intervened) was to force them into more market-oriented behaviour than before. Hitler in August 1939 spoke to his assembled generals of the 'devastating food crisis that could grip Germany within a few years', and may well have seen the Entailed Estates Law as a means of increasing food production.[78]

This brief moment in the sun helped Darré to pull himself together, and for a few months relations with Backe and his staff were harmonious. The constant notes issuing from his office were addressed to 'Dear Herbert': he visited the Backes at their farm, and stood Godfather to Backe's son. Hitler had expressed pleasure when the 1937-8 harvest was seen to be a good one, and told Darré that such a harvest was worth twenty-two divisions to him. Although the 'achievement', insofar as the improved harvest was a result of governmental activity, belonged more properly to Backe, Hitler tactfully applied the compliment to Darré, and indeed, as far as overseas visitors were concerned, Darré still had the title and dignity of minister. He was invited by Hitler to meet the Italian agricultural minister, and they planned in advance what to say to him. 'Refer him to Hungary for extra food', was Hitler's advice. 'We've got enough troubles of our own'. Darré received the Hungarian and Rumanian agricultural ministers in early 1939, while in July 1939 he organised his last International Congress of Agriculture held in Germany, and was flattered by the attendance of the Polish agricultural minister. Indeed, encouraged by his success with the Entailed Estates Law, he began to draft a new law in conjunction with Backe and Dr R. Harmening, the RNS's legal adviser, to create a new ministry which would combine the Agricultural and Economics Ministries into one vast planning and marketing organisation.[79]

Darré intended this proposed corporation to be the RNS writ large. By re-organising industrial production on the lines of a co-operative marketing system with fixed prices, agricultural price stability would be made easier, and industry gradually brought to see the virtues of corporatism. Backe agreed. He felt that the Four Year Plan council was failing to mobilise German's resources in sufficient quantity, and lacked organisational ability. More concentrated effort was needed, more efficiency and more productivity. One centralised planning department should replace the existing jumble of authorities. He could not see how Germany could win the forthcoming campaigns unless this disorganisation and sloth, as he described it after the war, was shaken off. The difference of attitude here was significant, with Darré thinking in terms of long-term reform, and Backe of the approaching conflict. In any case, unknown to Darré, the last props of his authority were being pulled out from under him. The one programme which was at the heart of his ideas—agricultural settlement—was threatened, and during 1939 the crucial transition of power to Himmler took place, which will be discussed in the next chapter.[80]

Backe responded to the prospect of conquest by reverting to his early Göttingen ideas about the need to colonise Russia's grain producing areas. He envisaged a restructured Europe, one unit from Brittany to the Urals, dominated by a strong Germany, and reasonably self-sufficient in food and raw material production. Russia would be Europe's 'bread-basket'. Both Backe's geo-political analysis and his devotion to the geographer and economist J. von Thünen (who developed a theory regarding efficient use of land resources based on closeness to markets) led him to disregard existing nation-state boundaries in planning the future Europe. Backe's study of Russia in the mid-1920s led him to predict the imminent collapse of the economy, and a decline in agricultural production. He rightly expected the Soviets to concentrate on industrial development rather than on improving agricultural infrastructure. He remarked that in the course of neglecting agriculture for industry Russian agriculture would be starved of resources, while industry could never develop in Russia because of the inadequacies of the population. Russian lack of creativity and initiative, together with Russian sado–masochistic attitudes to their rulers, going back to Ivan the Terrible, meant that Western-style industrial development could never succeed without the introduction of more creative and competent outsiders. The Germans, in fact, would act as an amalgam of colonists, managerial executives and studs. Russia deserved invasion, because her attempt to industrialise was tantamount to an act of war against the West. Her ordained role was to produce food for Europe. These arguments marched with Hitler's, and the fact that the renowned expert on world food production and on Russia propounded them, must have been a factor in his eventual invasion plans.[81]

In 1942, Backe was to have his thesis on Russian grain production finally published, along with a work on Europe's food balances. He wrote admiringly in 1943 that 'the Russian Bolsheviks cling on to their culture'. He was not a man to look for deliberate revenge for his treatment at the hands of the Russians between 1910 and 1918, but it certainly affected his view of them. None the less, his tough line seems to have been a result of his obsessive sense of identification with what he saw as German interests, a longing for large-scale, orderly planning, and a paternalist sense of duty, rather than a result of innate brutality. He had a strong instinct that the Germans were a defenceless, sacrificial entity ('We are the victims and not the creators of this world') who had to be protected, if necessary, through ruthless and revolutionary action. His Spartan ethic applied to himself as well as others. He refused to allow his wife to remove their belongings from Berlin during the worst of the bombing, because it would 'set a bad example to others', and refused her permission to travel to Berlin to see him in 1945 because the trains were 'needed for carrying refugees'. In fact, given the lack of apologies produced by most countries during their period of expansion or imperialism, an interesting question might to why so many German felt obliged to rationalise and excuse their aggressive policy during the war. Backe brooded throughout the war on an 'it's either them or us' line of argument, but he never faced up to the crucial misjudgment he had made regarding Russian food

producing capacities, a misjudgment that may have been due to the
fact that, like many Russian experts during the 1930s, who had left
Russia before 1920, at a time of relative plenty, he had very little idea
just how disastrous the effects of the collectivisation programme had
been.[82]

In January 1941, Backe produced a report on Russia's food
producing capacity which estimated a potential far greater than the
two million tons of grain Germany was receiving under the
Ribbentrop–Molotov Pact. Backe told Goering he was prepared to
guarantee the report's accuracy, and Goering showed it to Hitler, on
whom it apparently had a 'decisive effect'. On 30th May, Backe flew
to Obersalzberg to confer with Hitler, Lammers, Bormann and Keitel
(Darré was excluded) over the invasion plan and Backe's prediction that
'every country in Europe could share the *Kornkammer*'. The
responsibility for preparing food stocks and ration cards for the war
with Russia was given to Backe, who was under orders to keep the
whole thing secret from Darré ('row with Darré over secrecy of Russian
preparations', his wife noted.)[83]

Secrecy was now carried to considerable lengths as far as Darré was
concerned: clearly, he was no longer trusted. Bormann, especially, had
become an enemy. Darré despised Bormann's uncouthness, and his ill-
treatment of women. Bormann distrusted Darré, and especially his
commitment to organic farming, and, by implication, the
Anthroposophists. Even before the invasion of Poland, Darré had been
unable to gain Hitler's ear. His memos piled up unread on Hitler's
desk, but after 1939 his continual and open hostility produced not just
mockery from Goering and Goebbels but exclusion from political
life.[84]

One constant correspondent, Erich Dwinger, a best-selling 1920s
novelist who bought a farm in 1930 (the self-imposed silence of the
internal émigré did not prevent him from flooding the National Food
Estate with suggestions on farming and food policy throughout the
1930s), records a visit to Darré in April 1942, and described his
'somewhat too fleshy face, the eyes which had something in them of
the animal at bay, the well cut suit', and Darré's reiteration that he was
now a powerless man, and had been for some three years. Darré
confirmed to Dwinger that the Russian invasion was unnecessary for
Germany's food supply, but displayed instead 'naked imperialism'
which would 'finish off the peasantry'.[85] He was permitted to sit in his
office and keep the title of Minister, but it was on paper only. From
1939, Backe was described as acting head of the Ministry in its annual

budget. It took Darré two or three years to grasp his position fully. He kept up a sporadically friendly correspondence with Backe, warning him over Bormann in 1940, and sending him a long review of his own book, written in the friendliest of terms, in 1941. Backe, having achieved some autonomy at last, felt compassion towards him. After all, Darré had been his ideological mentor for some years, and on occasion Darré's concentration on far-reaching plans still seemed admirable rather than ridiculous. He commented that 'Goering is realistic but not good on fundamentals (*grundsätzlich*): Darré is the reverse', but none the less, blamed Darré for 'the fact that the peasantry were in an ever weaker situation' when faced with 'sarcasm over his pessimism and the peasantry in general' from Goering and Funk. Even as late as 1940, Hitler tried to pacify Backe over Darré, and called a meeting with the Ministry of Agriculture staff where, after congratulating Backe on his organisation of rationing, he asked him to compromise with Darré (who was currently campaigning against an agricultural study undertaken by the Ministry of Agriculture). Backe refused bluntly. He told Hitler that there was no point in trying to reconcile the conflicts caused by the virtual incorporation of the RNS National Food Estate into the Ministry of Agriculture, and Darré's resentment of the fact.[86]

When Goering issued a decree on 13th January 1941, concerning meat rations, the fact was kept from Darré until well into February. An angry correspondence with Backe ensued. Darré continued to circulate indignant memos and notes, and in June 1941 another row erupted, when in front of other staff he accused Backe of secrecy over the Russian preparations. Backe answered caustically that bureaucratic formalities were unimportant, and only realities mattered, a heavy blow at a man who equally disliked bureaucratic formalities, but, having been their victim, had tried to manipulate them against others. He refused to sign the rationing decree issued in April 1941. Backe reluctantly visited him in person to try to get the necessary signature, and Darré took the opportunity to try to explain his point of view. He complained that his task had been to rescue the peasantry, not to prepare them for war. 1933 had been a year of victory for the National Food Estate, but the Führer had diverted the nation's energies into building *Autobahnen* and other rubbish. Blood and Soil was a lost cause, but so was the war; the concessions made to Russia in the 1939 Pact had lost Germany the war with Russia before it started. 'He looked terrible, nose red, reading and muttering' noted Backe in an angry outburst. 'What he does not realise is that it is he who is shattered, not his work ... still, it's my problem now.' He went on to attack Darré harshly as a gambler and schemer, a worthless man who—and this was obviously the key to the attack—was trying to 'ascribe his own inadequacies to the Führer ... that Genius. I feel an unspeakable contempt for him', concluded Backe. 'To attack Hitler ... I myself am only a talent. I am happy to be a second-in-command, to work without personal ambition for the greatest .... In that I am completely Prussian.'

The very next morning Backe clarified his position still further, by describing a meeting with Heydrich and Landfried on the problems of raw materials. 'I was filled with joy to find that Heydrich was fighting for Nazi ideals. We were bound together, without a word being spoken.' But Himmler would 'collapse just like Darré, within three years at the latest; no accident that they were friends for so long', a prophecy that was to be fulfilled almost to the day.[87]

The link with Heydrich was not fortuitous. Backe felt a strong sense of community with him. He considered that they both suffered from capricious, vacillating, inefficient superiors in the shape of Darré and Himmler, a point also made by Heydrich's biographer. Both men placed their undoubted intellectual gifts at the service of a belief in planning as an aim in itself. The essential voidness of the concept, the impracticality of dictatorial technocracy, was obscured by the energy of the commitment. Backe's admiration of and comradeship towards Heydrich also reflects his sense that 'Nazi ideals' were lacking in their immediate peers. Two years later he wrote with regretful admiration of Franz Hayler, then *Leiter des Einzelhandels*, that he was 'Nazi and young and unbroken'. This suggests that 'unbroken' Nazi radicals were becoming thin on the ground. Shortly before Heydrich's assassination he was charged with investigating Backe, because of the latter's link with the Skald, and is supposed to have sent a message to Backe on his deathbed to the effect that he had cleared him of any disloyalty. This meant a great deal to Backe.[88]

Darré in 1940-1 attempted to fight back where he felt the seriousness of a situation demanded it, but lacked the nervous strength to put his case convincingly. He suspected that Backe was wildly overstating the figures on prospective grain production, and kept badgering the Ministry of Agriculture staff for clarification on this issue. He realised from bitter experience that Backe believed his own propaganda, the statistics which Darré feared were unreliable. 'Figures given to Goering as "political" figures were somehow transformed into "statistical" figures', he noted bewildered in August 1941. In 1939 and in 1940, he had sent reports to Hitler alleging that Backe had falsified food reserve figures, and managed to reach Brückner, an adjutant to Hitler, in 1940, attacking both the Four Year Plan prognosis of food production and Backe's prediction of food production in Russia and Germany. It seems true that Backe was prepared to bend the truth if necessary to convince Goering of the rightness of his plans. He was working all hours on the task of estimating Germany's food stocks and choosing National Food Estate personnel to take over Ukrainian collective farms. He feared that the effect of the invasion had not been worked out in terms of its effect on Germany's arms production potential.[89]

The food situation in Russia was to be a disaster from Germany's point of view. Backe had little idea of how badly forcible collectivisation had affected conditions in the Ukraine, or that the grain sent to Germany by Russia in 1939 and 1940 had had to be squeezed out of a near starving peasantry by force. Now, together with the damage they themselves had caused during the advance, the German armies found scorched earth—literally—and burnt crops. 'The further east, the worse the situation; nothing harvested, nothing in order—the Russian harvest uncertain, the Russians almost starving. In Germany, potatoes and sugar beet lying under the snow'. The vast grain reserves the Germans had expected to find in Russia were simply not there. By December 1941, Backe could note that 'the entire eastern army must be fed from Germany', at a time when the Ruhr potato crop had been ruined by frost. Backe's 'pragmatism' was shown up by this reversal of expectations. His policy over collective farms was that they had to be retained to avoid diminished food production in case the changeover caused chaos. It was a mistaken decision. Agricultural productivity did not pick up in the Ukraine until concessions were made to the peasants in 1942-3, but by then, the army was retreating, leaving huge quantities of tractors, gasohol plants and silos rusting on railway stations and on the steppes. If one is to judge from Russian reports of their endless agricultural disasters since the war, some, indeed, may be lying there today.[90]

These were the errors of an ideologue: a National Socialist totally committed to a credo which relied on the ethic of the *Volksgemeinschaft*. Backe, the archetypal National Socialist, underrated the speed with which people, even soil-bound farmers, even Russians, could respond to financial incentive. Since he envisaged a society fuelled by self-sacrifice and duty to the community, it is not surprising.

Darré was formally demoted in 1942 as Minister of Agriculture, and

Backe took over as 'leader of the ministry'. Remaining true to his Spartan values, he refused the title of 'Minister' until 1944. Both men had been uprooted and declassé. Both had unused and wasted abilities. In a time full of ironies, it is perhaps one of the deepest that Darré, the vain, difficult, unco-operative, intriguing politician, should have stuck to his beliefs, and that his prescriptions should so often have been the right ones, given Germany's position. Backe, the more upright, honourable and straightforward man, presided over the mass destruction of German agriculture, and witnessed the looting and expropriation of all farmers, large and small, by the Russian invaders he had always feared, in 1945. Trusted by Hitler to the last, he was nominated to the Dönitz cabinet of April 1945. In 1947, he committed suicide after several months of solitary confinement. A letter which reached him shortly before his death told him of the sufferings in the eastern zone of some of his closest friends under the Russians. It must have seemed 1917 all over again.

Darré, the stubborn individualist, went on to fight his corner till his death in 1953. In politics, as in art, ends and means, the virtue of the performer and the virtue of the end result, are strangely unrelated.

CHAPTER SIX

'The Praetorian Guard led by a Jesuit'

WAS DARRÉ GUILTY?

Four years after the war, in what became known as the Wilhelmstrasse trial, Darré was found guilty of the plunder, spoliation, enserfment and expropriation of 'hundreds of thousands of Polish and Jewish farmers'. In view of the fact that he was Minister of Agriculture in name only during the relevant period, was this verdict a miscarriage of justice?

In chapter Three, the concept of Blood and Soil was discussed in relation to intra-German racialism. It was argued there that Darré's concern was with rescuing a threatened minority from decay and possible extinction. Nonetheless, Darré seems to have presided, if in a titular fashion, over the resettlement of the ethnic Germans, the *Volksdeutsche*, in western Poland. He *was* the formal head of the ministry until 1942, and it was men from that ministry and from the National Food Estate (now incorporated into the ministry) who were sent out as civilian administrators under the military and SS aegis of the Wehrmacht and Himmler, to aid the Resettlement Corps in Poland. Furthermore, Darré had been head of the sinister-sounding SS Race and Settlement Head Office until 1938, and took a close interest in it during the 1930s, even soliciting funds for it in 1932 to fulfil his 'new nobility' visions. Although Darré tried to resign from the SS in 1938, but was prevented from doing so by Hitler's order, it is still a reproach against Darréan ideology that his visions of a peasant nation led inexorably to imperial plunder and land seizure; that agrarian fervour 'started Green and ended Bloody-red', in Tucholsky's phrase, and that the close tie with the SS seems to confirm this allegation.

The following two chapters discuss this issue. The first looks at the inception and structure of the SS Race and Settlement Office, and argues that it was at first a cardboard edifice, not nearly as important as its name suggests, and that early SS settlement was peripheral, a matter of a few part-time volunteers trying to sneak men and resources past the National Food Estate Settlement Office, and helping to build haystacks in Frisia. SS settlement outside Germany was from 1938, and took place without Darré's consent and above his head, after a series of complex manoeuvrings. An important turning-point in the power of the SS was the creation of the first SS Land Office in Prague, so while these institutional details are regrettably Byzantine, they are essential to an understanding of the underlying issues.

Chapter Seven looks at the details of ethnic German resettlement, and examines the extent to which Darré and his ideology were relevant to this improvised programme. During the early part of the war, plans were appearing and disappearing like the Cheshire Cat in *Alice in Wonderland*, and internal arguments about competence were proceeding among different SS offices, with Himmler holding the ring between them. It does emerge that some part of Darré's ideas had percolated through to the SS managerial level, especially where the social reformist, well educated intellectuals were concerned, but it seems that this process was indirect, and involved a rejection of what was seen as outdated romanticism. What came to be at stake was the belief, simple and non-ideological, that truly to possess a territory, it had to be settled by the possessors. Twentieth-century history amply demonstrates that this belief was not confined to the SS. So here, as in the question of agricultural productivity, it was the practical, populist element in Darré's ideology that survived.

The basic difference between Darré and Himmler was that Darré was a racial tribalist, and Himmler an imperialist with romantic racial overtones. Darré took seriously his own analysis that a nation could only be built upon an organic link between soil and *Volk*, but Himmler dreamed of a Greater German empire. While Darré had originated the *Hegehof* idea (later to be modified into Himmler's *Wehrbauern* concept), he had intended it as the basis for restructuring German society, not as a means of creating a ruling class for an Empire, or rewarding successful generals.[1]

Darré had to go, not only because he lacked the ruthlessness to implement Himmler's plans, but because he had come to realise that his dream of a peasant nation had little to do with Himmler's plans for the SS, the 'Praetorian Guard under Jesuit leadership', as he described it in 1938.[2] In some ways, the SS-State envisaged by Himmler had also moved from the original image of the NSDAP. Hitler, especially in his speeches, had presented National Socialism as the creation of the Greater Germany, purged of its alien and treacherous elements, a vision which could be perceived as an extension and fulfilment of nineteenth-century nationalism. This was what seems to have attracted the support of the majority of Germans.[3] Himmler's vision of the pan–European state, with a German and Germanised administrative elite, demanded the creation of a new kind of man to rule it. The ideal SS–man would be linked with the land, but would understand technology, and have mastered it. He would be efficient, able and unhampered by ties of class. This plan to create an élitist superman, which in many people's minds today is synonymous with National Socialism, was arguably different in kind from the cosy *Volksgemeinschaft* image presented by the NSDAP during its early years. It is interesting that even as late as 1940, Himmler thought it necessary to keep his plans a secret. Where the NSDAP's emphasis on community feeling exposed its real lack in Germany, Himmler's training and organisational methods managed within a few years to create a tightly knit and loyal group of skilled technocrats, together with a disciplined and skilful fighting force in the mature Waffen SS.

Between 1936 and 1939, the SS's position in agriculture was weak, even though SS indoctrination emphasised the peasantry and rural life.[4] The 1939 plan to resettle the *Volksdeutsche*[5] in the newly incorporated areas of the Warthegau, Posen, Danzig–West Prussia, Upper Silesia and Zichenau produced major alterations in the agrarian administrative structure, as detailed arrangements to settle the returning ethnic Germans took precedence over internal agrarian settlement. During 1940, a mere 619 new holdings were established within Germany, whereas some 35,000 farms were created for the *Volksdeutsche*.[6]

Whereas the Sudetenland appears to have been absorbed by German agricultural institutions without significant change, the war in Poland in 1939 produced a different order of priorities into the thinking of ambitious and/or competent officials. Konrad Meyer, ex-Ministry of Agriculture official and land planning expert, who took over Himmler's land planning office in Berlin for the *Reichskommissariat für die Festigung deutschen Volkstums* (RKFDV), indicated this conflict and the agrarian lobby's awareness of their loss of status at the end of the peacetime period, when he describes how RNS officials became demoralised as

> the static model of Darré's peasant policy became deeply affected
> by industrial and economic growth. Only defensive answers
> were given to critical questions. A resigned attitude reigned.[7]

At the same time, the SS were growing visibly in power and influence, and as a result, many members of the agrarian departments joined Himmler's new commissariat, the RKFDV, hoping to find there the same values that had inspired Darré. Himmler also recruited intellectuals of the traditional German nationalist variety, such as Dr Kummer, head of the Ministry of Agriculture Settlement Department, and ex-adviser to the Society for Internal Colonisation, together with Meyer; men who aimed at 'restoring the pre-1918 status quo'.[8] For them, the *Volksdeutsche* re-settlement was a matter of terminating what the Versailles Treaty had begun, sorting out the twenty-three major linguistic groups stretching from the Baltic to Greece.

Securing German expansion to the East had been a red thread in German politics; during the First World War, Max Sering, for example, widely regarded as a liberal intellectual, and certainly not a Nazi supporter in later years, produced a detailed plan in 1915 to establish 250,000 German peasant settlements in Courland. Arrangements were made with Baltic landowners to give up a third of their land for this purpose after the war. The loss of what became Lithuania after 1918 ended the programme.[9] In 1917, a *Vereinigung für deutsche Siedlung und Wanderung* aimed at settling German peasants from the Russian interior in the Baltic lands then occupied by Germany. Dr Stumpfe, a member of the Prussian Agricultural Ministry, suggested 'solving the German-Polish problem' by an exchange of population: Poles in Germany against Germans in what was then 'Russian Poland'.[10] There was apparently little difference between the earlier plans to repopulate the east with German peasants and Himmler's vaguely worded proposals of 1939 and 1940, with the exception of the proposed division of Poland into the *General-Gouvernement* (G-G), a sort of racial dustbin, and the Incorporated Areas, which were to be German, Germanic and/or Germanised.[11] This appearance of continuity has to be borne in mind if one is to understand the enthusiasm with which sincere followers of Darré, the peasant radical, altered course to follow Himmler the imperialist, and further, the extent to which Himmler was subjected to pressure from his staff on the issue of German and *Volksdeutsche* settlement.

Many of the institutional conflicts continued between SS Offices, party officials, Goering's Four Year Plan Council, and the OKW,[12] while disagreements about relative productivity levels took on a new urgency under the demands of the war economy, and the need to feed troops as well as civilians from occupied territory. The SS agricultural planning staff took over not merely the personnel, but the controversies involved. One important point, however, is that serious

SS involvement in the settlement programme did not occur until 1939. The extent and depth of its activity in this area has been exaggerated in later histories, partly because of SS power during the war, and partly because of the apparent continuity signalled by the existence of the SS RuSHA during the 1930s, a body whose level of activity at that time did not correspond with its grandiose title.[13] Other bodies were concerned with the change in authority from the Ministry of Agriculture to SS organisations, notably the RKFDV, but also the Hitler Youth Land Service, the Berlin Land Office, the *Vo-Mi Stelle* and the Prague and Austrian Land Offices.

Koehl describes in his monograph on German resettlement policy the methods by which Himmler organised SS infiltration into *Volksdeutsche* agencies, to win control of ethnic German settlement, and to persuade Hitler that he was the most suitable man to control the programme.[14] This resulted in the Führer Decree of 7th October 1939, which established the *Reichskommissariat*,[15] and although the Decree did leave control of 'settlement by farmers' in the hands of the Ministry of Agriculture (possibly due to Herbert Backe's influence), Himmler's role as *Reichskommissar* gave him the chance, which he utilised to the full, of dominating German resettlement activity in eastern Europe.[16]

This decree saw a final breach in Darré's relations with Himmler. Up to 1936, the two men had been friends. Backe, in fact, blamed Darré for inciting a romantic streak in Himmler's character,[17] and the two men did share a common enthusiasm for pre-Christian Germanic religions, traditions and myths. However, Darré's growing alienation from the party leadership, and Himmler's tendency to range himself with the existing powers (anyone who could move from being Strasser's devoted secretary, to Darré, to organising the Sicherheitdienst and SIPO in eight years certainly demonstrated ideological flexibility), emphasised a profound gulf in the ideas of the two men. Between 1919 and 1931, agrarian ideology seems not to have touched Himmler's political life. He was to all intents and purposes the perfect bureaucrat.[18] The process of ousting Darré had begun in 1936, when Himmler decided that SS settlement should be an independent matter,[19] while 1938 began with pressure from Himmler on Darré to resign his position as head of the Race and Settlement Office (RuSHA) a step which Darré considered a 'wound in the flank of the peasant struggle, a disaster for the peasants'.[20] He resigned in February 1938. In March 1939, Pancke, under Himmler's orders, successfully ousted the Ministry of Agriculture from effective control of land in the

Sudetenland, the reason given being, interestingly enough, that a successful SS settlement programme there would lend weight to an SS takeover of the same programme in Germany.[21] In other words, as late as March 1939 the programme of agricultural settlement *within* Germany was still seen as a tempting prize, a mean of creating an SS, land-based élite.

Where most Nazis dreamed simply of revising the German border to what it had been before 1918, Himmler had greater plans. Not only were five to six million Poles and Jews to be evacuated to *Restpolen*, not only were the *Volksdeutsche* to be resettled in the incorporated areas, but the whole of the east was to become a substitute to Germany for her lost colonies. In a speech in October 1940, Himmler attacked what he called 'ivory tower concepts of colonies, of open doors and new homelands'. Instead, the east was to be an economic area, a source of raw materials to be developed according to Germany's needs. After the war, new cities, industries and farms would be built under SS leadership.[22]

In trying to trace the genesis of Himmler's later career, historians have struggled with a paucity of material; there are early diaries, two brief articles published before 1933, an undated article on farming, and the probable part-authorship of the NSDAP Manifesto on Agriculture of 1930. One diary entry by Himmler does refer to possible emigration to the East,[23] and some writers have tried to see a continuity between this entry, his alleged membership of the *Artamanen* group in 1931, and the planned SS state of 1940. This interpretation suggests that Himmler, having been strongly influenced by Darré in the early 1930s, then proceeded to carry out Darré's programme, while discarding Darré the man.[24] However, an analysis of the planning documents and organisational struggles within the SS between 1938 and 1942 demonstrate that while agrarian ideology of a Darréan nature strongly influenced SS members at middle and lower levels, their attempts to implement peasant settlement were frequently delayed and even halted by Himmler, because Himmler was set on the potential empire in Russia, and subordinated ideological aims to power struggles. As with Darré, Himmler wanted agriculture to play an important role after the war; but his main purpose, to alter the demographic, political and geographic map of Europe, should be differentiated from Darré's Germanic *Bauernreich*. Himmler's plan was to develop the east into a mightly industrial empire, to re-afforest the steppes and mine the raw materials, using a helot class of Poles and Jews, tucked away in Russia. Himmler's wartime planning documents, while not of very great value in determining the course of events, are certainly of value in determining his intentions. We know from his subordinates that all such documents had to have his line-by-line consent.[25] They demonstrate that apart from the stress on the desirability of a medium-sized farm structure—and even this point could be dropped by Himmler on the whim of the OKW—the motivation of the two men was entirely different, a difference as great as the difference between Cobbett and Joseph Chamberlain.

Indeed, it can be argued that Himmler's plans for exploiting eastern Europe and Russia, far from being a direct result of the agrarian ideology of the 1920s and 1930s, was in many ways its opposite, and went further than the political expansionism that characterised Germany from 1890 onwards. It was imperial and not nationalist; technocratic instead of organic; authoritarian, where Darré sought, however unrealistically, to unwind state power.

Not all these subtleties were clear to Darré's subordinates. Indeed, Gauleiter Eggeling, formerly Keeper of the Seal for the German Peasant Council, wrote to Himmler in 1940, asking him to back Darré, because the farming population had become increasingly disaffected, and saw in Darré's eclipse a symbol of their own.[26] Furthermore, Himmler had developed a considerable admiration for the practical cast of mind of the RNS experts, and preferred dealing with them to the ministerial civil service. Darré in the meantime withdrew to his campaigns on organic farming, and his Society of Friends of the German Peasantry (a group he started in 1939, with a view to protecting peasant interests from the implications of the war, and the defeat which he foresaw),[27] while intermittently attacking Himmler and the establishment of the RKFDV.[28] Ironically enough, one of the most important methods used by Himmler to take over much of the Darréan *mythos* was the Race and Settlement Office founded by them both in 1934, and a discussion of the gradual vivifying of this cardboard edifice may help to explain some of the later complexities of loyalty that arose in 1939 among Darré's followers.

EARLY SS SETTLEMENT AND THE SS RACE AND SETTLEMENT OFFICE

Although the basic tenet of the Race & Settlement Office lay in the SS Marriage Order of 31st December 1931[29] through which permission had to be obtained by SS members for marriage, and the prospective spouse racially examined by the SS Race Office, it was not until

Himmler's decree of 21st September 1934 that the office was effectively established as an independent vehicle for the ideological education of the SS.[30] It was to bring about a close connection with the peasantry, and realise the aims of *Blut und Boden*.[31] Although the peasant theme ran through the office, the Settlement Division had the lowest but one number of officials manning it: two, as opposed to the ninety SS men working in the *Sippenamt*.[32] This indicates the lack of importance accorded to SS settlement; none the less, the training courses for SS members included education in fundamental concepts of *Blut und Boden*[33] and the Darréan theme was emphasised in other ways. For example, each department·had a *Bauern* representative, who wore the Odal rune as an insignia.[34]

Darré was head of the RuSHA until February 1938, when SS Gruppenführer G. Pancke took over. In July 1940, Otto Hofmann, and in April 1943, Richard Hildebrandt, succeeded to the position.[35] The racial department had been established to propound the view that all history, ethics, law and economics are determined by blood', and worked largely on theoretical questions. It actually was run by Dr Reischle, second-in-command of the RNS, and greatly admired by Darré, whose biographer he was. In 1939 he was sacked by Himmler, courteously enough, for being a 'Darré man', and replaced by a Dr Schultz.[36] The Racial Office itself was hived off from RuSHA in 1938, and became part of the *Ahnenerbe* office.[37]

Werner Willikens, State Secretary in the Ministry for Agriculture, and a close friend of Herbert Backe, headed the settlement division, which was supposed to deal with settler selection, applications and homestead settlement; all of these were matters which in fact were dealt with by the Ministry of Agriculture settlement department. In 1936, Backe took over from Willikens, and three new departments were created, but no significant change in activity occurred. In March 1941 the RuSHA took on the role of welfare office for the Waffen SS, especially those members who were of peasant origin, and later on in the war ran convalescent homes for wounded members of the Waffen SS.[38]

The fact that two prominent members of the Ministry of Agriculture, Backe and Willikens, were successively leaders of the settlement division of the RuSHA seems to emphasise the importance of this office. However, it must be stressed that the real activity was going on in the Ministry of Agriculture itself, and the positions of Backe, Willikens and Reischle emphasised RuSHA's weakness rather than its strength. They were there to lend it authority and status, their role largely confined to signing letters and bulletins.

The cross-membership existing between the Ministry of Agriculture, the National Food Estate and the SS, implies an identity of purpose which did not really exist. The upper echelons of the agrarian lobby, such as, for instance, all *Landesbauernführer*, had always been encouraged to join the SS, partly because of Darré's original connection with the formation of the Race & Settlement Office, and partly because of the early friendship between Himmler and Darré.[39]

During the war, the agricultural experts employed outside German territory were normally given SS ranks. Cross-membership in the SS continued well into the war, as indeed was inevitable, since no fully autonomous SS units seem to have existed before that.[40] There was a certain amount of cross-membership in the SA, including Gustav Behrens, peasant representative in the RNS. But in general such membership was discouraged. Darré disliked the SA, considering them to be largely urban in composition, stupid and overbearing.[41]

The first SS settlement project was in 1935, the *Gemeinschaft der SS—Siedler und der Nordensee*. It soon encountered difficulties, largely because the SS-men who were building the settlement were doing so in their spare time. It was at this time that Himmler complained to Harmening, of the RNS settlement department, that settlers were divided into 'pure organisers and romantics'.[42] Despite this bad start, Himmler wrote the following year to Darré, stressing that details were not as important as precipitating action.[43] Darré's odd handling of Himmler's request was connected with his growing fear of Himmler's power. Aware of his own weak position, and Himmler's interest in taking over settlement activity, Darré had tried to engage Bormann, of all people, to support him against Himmler as early as 1935. However, after a certain amount of wining, dining and plotting, Bormann disclosed Darré's approach to Himmler,[44] who seems to have been amused rather than alarmed. The result was that Darré began to evade Himmler's letters and verbal requests about SS settlement from 1935 onwards.[45] But during 1936, a barrage of requests came to Himmler to make room for SS-men in settlement villages. These were passed on to Darré.[46] Clearly, although Himmler had always intended the RuSHA to establish SS settler communities, it was not a high-priority issue, and the principles of comradeship, mutual loyalty and support which his orders to the SS emphasised, conflicted visibly with the principle of economic viability.[47]

The exchange of letters with Darré—still 'Richard' and 'Heini' at this stage—suggests that Himmler's interest in SS settlement arose more from pressure from his staff than from some long hidden plan of his own. Himmler decided on a pilot scheme and the building of an SS settlement village, the first to be formally supported by the SS Head Office, was put in hand late in 1037. It indicates Himmler's growing interest that Dr Kummer demanded six-monthly reports and asked the local SS unit to help supply emergency hay and straw.[48]

This was the first indication of what was to be Himmler's way in to the settlement programme—confidential co-operation with Kummer, head of the Ministry of Agriculture Settlement Office. It was at the same time, late in 1937, that the Race & Settlement Office, after hearing from Himmler, announced to Kummer that they wanted to see SS settlement increased, and suggested Pomerania. Himmler agreed enthusiastically, promising that land would be made available for the SS from the amended *Osthilfe* procedure, whereby indebted estates exchanged land in return for government subsidies. Kummer, of course, was not supposed to syphon off settlement land and funds to the SS; either nobody noticed what was happening, or they were too frightened of Himmler to complain. Kummer also gave detailed instructions to the SS section leaders on how to avoid conflict with other settlement offices, together with the NSDAP at *Gau* level.[49]

Writing to Himmler, Kummer spoke even more frankly about potential institutional conflicts, indicating the extent to which the SS was distrusted by other groups within Germany in the 1930s. Even Darré was alarmed at news of the 'Fausthof' project. The first result of the publicity which surrounded its founding was that he indignantly demanded details of it, including the relevant cost estimates. Kummer kept Himmler advised of Darré's reactions, and suggested that.

> In my capacity as a member of the SS and not as a civil servant, the Northern SS section should at once organise their own SS activity, since close co-operation between the settlement authorities and the SS is lacking.[50]

These authorities had a high proportion of pre-Nazi civil servants, and possibly for that reason were more hostile to the SS. Certainly, Kummer alleged that these same authorities were 'ganging up' against the SS, together with *Landesbauernführer* from the National Food Estate, and that both groups had Darré's support.

Land acquisition for those SS projects encountered the same difficulties that faced the Ministry of Agriculture. Domaine land was seldom available for settlement, while the OKW supported the existence of large estates on the grounds that they were the only effective farms.

This discussion of early SS settlement activity brings out several points which have not emerged clearly in the literature concerning either the SS and agriculture, or the SS's relationship with other institutions between 1933 and 1939: first, that SS settlement was a late phenomenon in the Third Reich, despite the emphasis in its own propaganda on *Blut und Boden*; second, that the head of settlement in the RMEL was plotting with the head of the SS to bypass his own ministry, and specifically the man who had propounded the notion of settlement in the first place, Darré. It demonstrates the degree of hostility existing between the National Food Estate and the SS, despite some degree of interlocking membership at top administrative levels,[51] and the extent to which the Settlement Office and peasant education department of the RuSHA remained a sideline until 1937. Kummer saw no difficulty in working towards what he assumed to be Himmler's aims, having always been a keen German expansionist, obsessed with the need to expand and strengthen the German border in the east. He promised Himmler land for the SS via *Osthilfe*, in order to 'create important racial and political support for the Führer's policies *east* of the Elbe'.[52] Indeed, Kummer, chief financial adviser to the *Gesellschaft zur Förderung der Inneren Kolonisation* for many years, was writing to Darré as early as 1933, that settlement in eastern and northern Germany had to serve the national interest and 'maintain the armed political situation'.

This implied a complete reversion to Wilhelmine and Weimar concepts of eastern settlement, and a final rejection for Darré's *Bauernreich*, and helps to demonstrate the continuity that lay behind what became a key SS programme, poorly carried out and unfinished as it was: the establishment of a *Volksdeutsche* barrier to possible dangers from the east. The value to Himmler of the SS settlement programme in Germany seems to have been a pilot project. Himmler himself referred to the need for a 'copybook exercise' in settlement in late 1938, for the RuSHA to obtain the 'foundation and experience for a future huge settlement proceeding', and although he did not state where the future proceeding would take place, it was presumably to do with the ethnic Germans.[53]

Among ordinary SS members, however, there was enthusiasm for settlement on German territory, although most SS members were employed in non-agricultural posts. In December 1937 all leaders of the Death's Head Division were asked to find out how many of their members were suitable settler material, and how many members came from peasant backgrounds.[54] Interestingly, the collation of these simple statistics was ordered to be kept secret 'to prevent unrest among units'; presumably, settlements were so desirable that jealousy and competition had to be avoided.[55]

Another initiative which emerged directly from local SS section leaders showed again their interest in increasing the SS link with the land. The Saxony *Landesbauernführer* agreed with the local SS leader to form SS agricultural units, which would work with the local RNS members in helping to organise settlements.[56] Darré forwarded the idea to Himmler, who had no objections.

Despite the apparent co-operation of the *Gauleitungen* in Saxony and Swabia, who offered no opposition to the SS recruitment drive among young people for the agricultural units, activity in this area continued to be merely marginal and local.[57] Yet, if ideological 'correctness' was any guide to success in the Third Reich, such projects should have been more effective, for the ideological reasoning behind the formation of the SS/*Reichnährstand* groups was strongly emphasised in the recruitment literature. One directive stated:

> The SS is part of the NSDAP. Its members are increasingly selected with a view to their racial value. Their racial excellence can only be perpetuated if the SS is rooted in the peasantry. Herein lies the deeper meaning of the concept *Blut und Boden*: all young men from the countryside should be SS members. The SS *Landgruppen* will be the future racial crack-troops of the peasantry.[58]

In reality, evidence suggests that the link between countryside and SS-men was less significant than the propaganda suggests. Statistics conflict on exactly how many SS-men came from peasant stock, but among the long-serving SS-men who had received farms by 1941 many had done badly through lack of experience.[59] The fact that their wives were usually from an urban background was an added drawback. This lack of success was despite a compulsory two-year training course.[60]

When the right candidate was available,[61] then (despite grand claims that shortage of money must not be allowed to deter applicants), money for SS settlement was a major issue, as it was for the RMEL settlement department. In August 1938, Pancke, who was now head of the RuSHA in Darré's place, warned Himmler that the RuSHA's choice of suitable settlers was being curtailed by 'a shortage of money among settlers'. Pancke conferred with Walter Granzow, an ex-RNS official who had been promoted to be head of the *Deutsche Siedlungsbank* after a failed attempt to oust Darré,[62] and Granzow agreed to provide bank funds of three to eight million Reichsmark.[63] He refused requests for more on the grounds that some ten thousand settlements had been in financial difficulties in the last year alone, and had received six million Reichsmark.

After the creation of a Protectorate in 1938, the RuSHA looked to the Sudetenland to provide a supply of cheap, if not free land for settlement. It tooks the fight for SS settlements into a new area and brought new room for manoeuvre, plotting and power-seeking, all areas in which the SS agrarian ideologues proved adept.

THE LAND OFFICE IN PRAGUE: SS INSTRUMENT FOR EXPROPRIATION

SS settlement activity found its suitable vehicle in the shape of the *Deutsche Ansiedlungsgesellschaft* (DAG), or German Resettlement Society, the largest of the settlement societies re-organised by Darré between 1933 and 1936. In 1939, the company was still in private hands, and when its manager suggested to Darré that the shares should be nationalised, Darré agreed, seeing the proposal as a move away from 'capitalist concepts'. He ordered the shares to be sold to the Dresdner Bank as a beginning. He was unaware that von Gottberg, an official in the RuSHA, had secretly told Riecke, Backe's second-in-command in the Ministry of Agriculture after Backe replaced Darré in 1942, that he was interested in purchasing the shares on behalf of the SS.[64] When the purchase was completed, the Dresdner Bank told Darré that the two government officials who had bought the shares were Gottberg's nominees. Gottberg now controlled the DAG. Darré protested to Himmler that he had disbanded the earlier settlement societies because of their 'strongly capitalist orientation', and that, if a single SS unit controlled the company, the interest of Germany and of the settlers would be unprotected. On Himmler's order, von Gottberg submitted a reply which, interlarded with the standard jargon about the need for a healthy peasantry and *völkisch* economic policies, invoked Wehrmacht support and contained the delicate threat. 'Nobody is likely to take responsibility for destroying such vital measures'.[65] In other words, the whole exercise was in order to establish SS farmers in Germany where uprooted by the Wehrmacht, and sidetracking the hostility that would otherwise be encountered. The share purchase is dated at some fifteen months before the date on which the DAG was alleged at the

Nuremberg trials to have been acquired for the purposes of expropriating land in the East.[66] In spite of Himmler's dislike, the management of the DAG was left strictly alone, and proved the ideal instrument for organising large-scale settlement of ethnic German peasants in Lorraine, Danzig-West Prussia, the Tyrol and the Danubian area, and by 1940 was being used for this purpose by the SS.[67]

In 1937, the SS had formed a non-profit-making corporation to buy up church land in west Germany standing empty as a result of the closure of the Catholic Orders by the state. The society, the *Deutsche Reichsverein für Siedlungspflege*, or DRV, was registered in Berlin, where it bought shares in the DAG with the help of the Prague Land Office. Hildebrandt, Pancke and von Gottberg were on the Board of Management, while Theo Gross, first head of the Prague Land Office, was liaison officer with the NSDAP. Although registered as an ordinary share company, its articles stated that it was not an 'economic commercial enterprise', but looked after the welfare of all German settlers, coordinating the available facilities of various German counties 'according to the National Socialist *Weltanschauung*'. This in effect meant a potent anti-church bias, that came into full force in south and west Germany, as well as in Austria, and the Protectorate with its strong Catholic church. As their 1940 report pointed out, 'The ideological fight against the political power of the church takes a new turn with this instrument' (the DRV), which claimed to be merely returning to the German people property expropriated over hundreds of years by the church.[68] This legalised take-over was carried out in association with the Prague Land Office. What had begun as a means of getting over the shortage of money and resources for German based settlement had become a large-scale instrument of resettlement and expropriation, carried out with a complementary paraphernalia of welfare organisations and social workers. It was perhaps a case where ideological and financial requirements were happily married.

After the Anschluss, the new governor of Austria, Seyss-Inquart, suggested to Darré in May 1939 that a special office, staffed by the RNS, be set up to deal with the special problems of the Austrian mountain peasants.[69] These peasants' poverty had shocked visiting German agriculturalists in the mid-1930s.[70] Now Seyss-Inquart wanted immediate aid for them while the dissolution of the Austrian Ministry of Agriculture meant that administration now devolved upon the German ministry. The southern peasants conformed to the National Socialist agrarian ideal, being self-sufficient, producing little marketable surplus and 'of great racial value'. Darré agreed to co-operate in setting up a separate office in the area, adding that he would be particularly glad to introduce into his ministry someone who would 'redress its somewhat strong north German emphasis' (a reference to the effect of the incorporation of the Prussian Ministry of Agriculture in 1935).[71]

When Darré put the idea before Lammers. it was refused, as was usual with Darré's proposals.[72] However, Heydrich heard of it, and forwarded a suggestion to Himmler that a Land Office be established to manage land both in Austria and in the Sudetenland, to be based in Prague. The Land Office was duly established in July 1939, on the face of it a joint effort between RNS officials and local representatives of the German minority.[73] Its function was to oversee the evacuation of Czechs from German and Germans from Czech areas of the annexed portions of Czechoslovakia.[74]

Unknown to Darré, the plan for the SS to control the Sudetenland via a Land Office had been proposed by Pancke to Himmler in October 1938, in a letter discussing the possible expansion of German settlement there.[75] After Henlein's nomination as *Reichskommissar* for the Sudetenland, Pancke decided to ignore Himmler's previous order that there was to be no contact between the ethnic Germans there and German officials, and held a meeting with Henlein and Karl Franck on 'how to rectify the results of the Czech land reform' (a reference to the anti-German legislation embodied in the Czech Land Reform of the 'thirties), and establish new German settlements in the 'Sudetengau'. Henlein responded at a later meeting that the Sudeten German peasants wanted to have nothing to do with the SS—exactly why was not stated—and Pancke reported to Himmler that the best way to insert SS influence was to set up a Land Office, with Henlein as leader, and RuSHA men providing the personnel, to 'maintain SS influence'.

> If, through the work of the SS and RuS members in the Sudeten area we can achieve an example of correct and healthy settlement, then the moment will be brought nearer when the SS, under the spur of true achievement, will be able to take over the Settlement Office in the old Reich as well ... the work of the Settlement Office [of the RuSHA] will remain trivial unless the whole area is decisively altered, an opportunity which now offers itself in the Sudetenland—either that or spend millions.[76]

—presumably a reference to the possibilities of seizure without compensation of evacuated land, or land under the control of the Gestapo.

Parcels of land were seized from Czechs in the Sudetenland by the Gestapo, the only area in 'Greater Germany' where this procedure seems to have been followed (although the RNS LBF for the Saarland reported mass evacuations from land holdings). Acquisitions in the Sudetenland averaged 520 Reichsmark per hectare, whereas compensation prices in Thuringia, for example, were 2,600 RM per hectare.

A puzzled reference by Lammers to Pancke's draft suggestion for the Prague Land Office survives. In the same month that Pancke wrote to Himmler, October 1938, Lammers noted that a member of Henlein's staff in the Sudetenland had negotiated a draft *Führer*-decree with someone in the Chancellery on the regulation of land use in the Sudetenland, proposing the erection of a Land and Settlement Office which would be completely independent of the central authorities, with the exception of the Four Year Plan authorities. Lammers minuted that the Minister of the Interior thought the proposals 'impossible', while none of the departments concerned could be found to admit knowledge of the matter.[77]

Himmler appeared to favour any plan put on him for extending SS authority—as long as it could be done without too much friction. The establishment of the Prague Land Office in 1938 was important. It was the first example of SS institutional imperialism at work outside the *Altreich* borders. Darré, the RMEL and the RNS were all successfully bypassed, and the Ministry of the Interior's objections ignored. The success of a small and hitherto unimportant SS office in gaining control in the occupied territory was a step towards collecting all the racial offices under Himmler's control, a *fait accompli* by the end of 1939.[78] It gave scope for the first massive transfers of land, property and people.

Darré, meanwhile, had attempted to resign from the SS—a brave move blocked by Hitler on appeal by Himmler[79]—and with his gift for acquiring the hostility of influential opponents by inept political man-oeuvring, had begun a campaign against von Gottberg, the official behind the SS takeover of the DAG. Complaining as he would later of 'SS Cheka methods used in the Sudetenland against Germans',[80] Darré wrote

SS Führer von Gottberg, [now] leader of the SS Land Office in Prague. A drunken sot involved with gangsters. The evacuation policy in the Czech area, yes, but slowly. Gottberg, quicker but worse. On this, it must be said that *my* ordinary administrative facilities could not have coped with this 'de-Czeching' of Bohemia and Moravia, because to do this 'legally' would have required extensive preparations over a long period of time, at least, as long as the German government valued appearances before the outside world.[81]

—a startling comment from someone who was still a minister of the government, and one which revealed again his alienation from what was happening.

In an attempt to further his campaign against von Gottberg, and at the same time rescue his own concept of peasant settlement, Darré contacted Gustav Pancke in May 1939 (at that date Pancke was still head of the RuSHA) to try to win him over. Darré's diary records only the discussion over the planned *Wehrbauern* villages on Germany's eastern border, but Pancke's notes, drawn up immediately after the meeting in Darré's house, show the attempted negotiation in detail.[82]

Darré first told Pancke that von Gottberg was trying to establish a Land Office in Prague over his head, and suggested that Kummer should replace Gottberg. Unfortunately for Darré, Kummer was now in disgrace with the RuSHA. He had criticised the DAG to a *Landesbauernführer*[83] and Pancke commented ironically that co-operation with Kummer would be difficult, since he attacked all Race and Settlement projects. Darré then turned to Himmler's *Wehrbauern* concept, which entailed a fortified village of some thirty families, governed by two or three SS men who would also be trained soldiers. Darré attacked this idea of combining farming and fighting, and again propounded his own belief in a peasantry closely rooted to their own land, drawing a parallel, once again, to the colonisation of the West in America.

Pancke did express agreement on the basic idea of strengthening peasant representation: 'I hope that one day every Minister and State Secretary will be a farmer's son', but he differed substantially on the rationale for settling 'relatively uncivilised areas in the East'. It was not to possess the land in some rooted, mystical way, but to form economic ties with the *Altreich* which would bind the two areas together irrevocably, as far as marketing, production and industry were concerned. He also doubted whether a pool of suitable peasant families existed for what Darré had in mind. Did such 'adventurous and romantic families really exist'? Darré's proposal for future co-operation became vague at this point, and Pancke retired in bewilderment, suspicious about Darré's intentions. Thus, Darré's attempt to win over Himmler's man on issues, both organisational and political, where there was considerable disagreement, had the opposite effect.

The shift of emphasis from establishing a large-scale SS settlement programme within Germany—the aim as expressed in Pancke's letters to Himmler in 1938—to creating what was eventually an SS ruled empire in Poland and Russia, was clearly foreshadowed in this conversation.

The invasion changed the focus of attention for many of the agrarian ideologues, as the example of Konrad Meyer was to show. Shortly after the creation of the RKFDV, Himmler summoned Meyer to Warsaw, and asked him to take over and expand the RKFDV planning office. Meyer, who had occupied academic posts before the war as an expert in land planning and utilisation, accepted.[84] He commented after the war that relations between his office and Darré remained tense until Darré's dismissal as minister in 1942, whereas 'relations between the RKFDV and the Ministry of Agriculture remained undisturbed ... through the sensible behaviour of State Secretary Backe'. Darré's public response to the news of the formation of the Reich Commisariat was to call a conference of representatives from the Ministry of Agriculture, the National Food Estate and the Settlement Societies to discuss their mutual expertise in agricultural settlement. He wrote in his diary: 'Harmening tells me that all settlement in Poland shall be carried out by the SS ... This is the most decisive defeat of my life.'[85]

Although much of the ensuing correspondence between Darré and Himmler, and Darré and Lammers, stressed questions of ministerial competence, and although the trial judge at Nuremberg considered that the disagreement between Darré and Himmler was over power rather than ideas, it is clear from Darré's campaign that issues of vital ideological importance for him were at stake.[86] He wrote to Lammers, for example, enclosing a book on Anglo-Irish history, saying that if the Polish resettlement was not carried out 'from the point of view of creating a sound land law', it would end with the same kind of strife that had characterised England's relations with southern Ireland, the north, according to Darré, having been stabilised through a yeoman-farmer pattern of development, while the south had been patterned on English neo-feudal land holdings, large estates and rented holdings.[87] This was a development of Darré's belief that independent farmers were more likely to identify with the nation, and less likely to behave as hostile nationals. The letter does leave unclear the extent to which Darré was prepared to dispossess the Polish farmers in the incorporated areas, beyond, that is, the question of exchange of populations, generally envisaged by Germans at that time. It is possible that he was thinking of a mixed German-Polish occupation, both farming according to the *Erbhof* ideal. Certainly Darré had never been as anti-Polish as many of his colleagues, as his dispositions for allowing Polish farmers to become *erbhoffähig* under the EHG make clear. However, what was more important to Darré than humane considerations about the existing Polish settlers was the need to avoid a colonial situation, the nightmare that had bedevilled England's relations with Ireland. He was also looking for a looser socio-political structure than was generally envisaged, one the Jeffersonian connotations of which are indicated in his comparison with the American colonisation of the West: the farming population was to be bound more to the soil than to national or economic entities. In contrast, the emphasis on technocracy, development and efficiency that developed in the SS-state (which is not to argue that the SS-state was efficient in practice), included the notion of a link with the soil because it fitted practical data, the given belief that social man needed such a link, and that a society ordered on such a basis would be more efficient.

The controversy over whether settlement was intended to take place within an enlarged German border[88] or whether it would consist of an armed incursion into lands east of Poland, obviously carried implications which went to the heart of Nazi ideology.

By 1938 Darré observed the growing power of Himmler and the SS with alarm—'Heini now has the soul of the SS firmly in his hand'. He was one of the first observers of SS power to comment on the growth of their *economic* power; he claimed in early 1939 that Himmler was deliberately infiltrating his men into positions where they could 'hold the purse strings'.[89]

The sub-stratum of policy disagreement was not a negligible element in Darré's attempt to protect his sphere of power, since he was convinced that a settlement programme carried out by the RNS—using the *Erbhof* legislation, and reshaping dwarf farms into viable units—would be different from and superior to SS-ruled expansion, with their emphasis on economic activity and, Darré believed, their hostility to the peasants.[90] When a draft *Ostmarkgesetz* appeared in April 1939, Darré was astonished to find that it would have taken all power, effectively, from the RNS, and he immediately complained to Lammers. Even Lammers saw to it that the draft was withdrawn. At issue was Himmler's attempt to convince Hitler that resettlement was a political matter rather than an agricultural one. To this end, the SS controlled the *Vo-Mi-Stelle*, and had achieved close links with several ethnic German communities abroad, particularly in Eastern Europe.[91]

Himmler's own plans for eastern settlement, surprisingly similar to the Weimar Internal Colonisation movement, concerned the dangers of Polish migration into border areas, and emphasised the notion of a border defended by militant peasants of pure German blood. Naturally, Darré also wanted the German borderlands to be as densely populated as possible with German peasants, but this was secondary to his main vision, and he could not be described as 'imperialist' in his aims. For example, he commented in March 1939, 'Protectorate declared: now possible to have a colony within our borders. Only bearable if the *Volk* feel a *Blutgedanken*. God knows what will happen'.[92] The whole dimension of the 'national interest' seemed to escape him, while Himmler's interest lay on the militant and expansionist role of the SS in the east.

Germany, unlike Poland and to a lesser extent Hungary (which both had *Wehrbauern* policies), was a highly industrialised and technologically advanced country which was currently planning an aggressive war; that Himmler should have been thinking in terms of Cossacks, Poles and feudalistic Hungarians, to promote a plan which implied a defensive military position and a reasonably stable border area (else why *settle* it?) demonstrates a curious unrealism. Certainly, *Wehrbauern* settlement was a very limited exercise, when it finally took place.[93] The failure of Himmler's plans fully justified Darré's criticisms of them to Gustav Pancke in May 1939, when Pancke, the new head of RuSHA, held talks with Kummer and Darré concerning the proposed *Wehrbauern*. After watching the peacetime settlement programme founder in Germany through a shortage of land and money, Darré was now presented with the full might of SS power being thrown behind a programme which he considered contrary to the ideals of *Blut und Boden*, and quite unrealisable anyway. He did not fight his campaign well. Prepared to develop the implications inherent in his own views—writing in June 1939 that the 'state and its spirit must be wound down, Megalopolis will decline'[94]—Darré entered into a condition of incoherent rage that his slogans should be used by men who intended something quite different.

On hearing that the task of resettling the ethnic Germans had been given to the newly established RKFDV[95] in October 1939, Darré wrote a long letter to Lammers claiming that the RNS alone had the special knowledge needed to resettle those ethnic Germans who had been expropriated by Polish land reforms. He insisted, with his usual tactlessness, that SS-men were the wrong choice to carry out the job, because they were particularly unpopular with the peasants, and any such connection would arouse resentment. There seems to have been little evidence for this claim, but Darré's next point was more valid: that the SS would not know how to create viable holdings from the 'dwarf holdings' common in the Warthegau. Darré's criticism of the *Wehrbauern* concept was then made openly for the first time. If the word meant simply a yeoman peasantry, able to defend themselves against bands of marauding Polish migrants, then the ordinary *Bauern* was perfectly suitable for the task. If a 'Cossack type of border protection' was at issue, then Darré demanded a clarification of what precisely Himmler thought he was playing at, since such settlement implied the existence of garrisons, and the superiority of such a military force over the farmer. In that case, the normal border garrisons and motorised troops would be more effective than the *Wehrbauern*, who were neither fish nor fowl,[96] too ill prepared for a modern war, and too involved in military relationships to develop into a stable peasantry.

Unfortunately for Darré, this sensible analysis appeared too late in the course of a long letter to repair the impression of a man merely fighting for his bailiwick. Darré's next step was to send a ten-page memo to Himmler about the background of the Baltic Germans from Estonia and Latvia who were expected 'any minute'.[97] Darré set out their history and told Himmler that he had laid a comprehensive plan for their resettlement before Stresemann in the 1920s. Himmler replied more than two weeks later, with a dig at Darré's expert knowledge: he alleged that the supposedly hundred per cent *bauernfähig* settlers from the Baltic in fact were mostly from the Baltic towns, while the remainder had acquired enormous latifundia through the old Hanse League (*sic*), and Himmler really did not feel able to reproduce the same conditions in the German area.[98]

Himmler was clearly not prepared to compromise with Darré, and it was only through Backe's intervention that some marketing corporation personnel were eventually 'borrowed' by the RKFDV and the *Vo-Mi-Stelle* for the task of organising ethnic German settlement, but by now under the effective authority of Backe and the Ministry of Agriculture rather than Darré.

Fundamental responsibility for the resettlement of the ethnic Germans lay with Himmler and the RKFDV, specifically with SS Brigade-leader Ulrich Greifelt, head of the RKFDV till 1942. Throughout the war, there had to be contacts with the Ministry of Agriculture over questions of ensuring harvesting, and transporting produce, so that while unconnected with resettlement, the Ministry of

Agriculture was involved in the administration of food production in the occupied areas. Regional farming advisers, known as *Kreislandwirten* were installed, who were responsible for ensuring food supplies locally, but had nothing to do with the question of land ownership.[99] Within the framework of the Four Year Plan, the *Haupttreuhandstelle Ost* was administered by Winckler (who had undertaken trustee work during the Weimar period). Darré's replacement by Backe in 1942 meant that a new phase of co-operation between the RKFDV and the Agricultural Ministry began. Backe made an effort to retain Himmler's favour, while endeavouring to maintain the level of food production in the incorporated areas; he may also have supplied an element of factual information for Himmler, who lacked a technical background in agriculture. Backe's promotion obviously simplified matters for Himmler, who no longer had to cope with Darré's tendency to play at machiavellian politics, while at the same time trying to salvage his own aims.

CHAPTER SEVEN

Poland: Germany's Ireland?

GRANARY OR PEASANT'S PARADISE?

In November 1939, Willikens, State Secretary in the Ministry of Agriculture, wrote to Himmler as *Reichskommissar*, suggesting that the function of the new *Reichsgaue* (the Wartheland) was to be for purely German peasant producers.

> No titles to the new land should be handed out, because after the war, priority must be to *Volksdeutsche* peasants and soldiers who fought for it. This is morally right, *also the way to produce as much as possible from the land.*[1]

Nothing could more clearly encapsulate the joint role the *Volksdeutsche* peasants were to play than this request. But a different plan was to use the conquered acres of the east to establish large estates on the Prussian pattern, to produce grain and potatoes with hired Polish labourers. This was favoured by Goering and the German High Command, while Himmler, despite his penchant for the *Wehrbauern* concept, favoured in practice the demands of the landed aristocracy.

The imminent arrival of hundreds of thousands of returning Germans turned the balance in favour of an attempt at a peasant-type settlement, at least as concerned the *Volksdeutsche*. Such an operation, however, necessitated the large-scale deportations which were associated with the re-settlement programme in the incorporated areas, and which have drawn attention away from other aspects of these events.

The existence of two quite separate plans for the area—the raw material-pool-cum-corn-chamber, and the Germanised peasant nation—was complicated by the improvised nature of everything that took place between 1939 and 1945. As plan succeeded plan, order succeeded order, and reorganisation succeeded reorganisation, great care has to be taken to distinguish what was actually happening in the incorporated areas from what the most recent pronouncement had demanded was to happen. For example, one might well wonder how it was that the *'Z'Aktion* of 1942 which expelled Polish dwarf farm smallholders from their holdings in the Warthegau, could find any Polish farmers in the area at all, considering that at least three orders had already emerged from various offices to deport them to the *General-Gouvernement*.[2] A year later, hundreds of thousands of Polish farm labourers were deported to Germany as conscripted farm labourers, workers who would not have been available for that purpose if previous deportation plans had been implemented.[3]

Even the exact whereabouts of the border between *Restpolen* and the new German county was a last-minute arrangement. One RuSHA adviser wrote scathingly to Himmler in October 1939 that the plan to incorporate Kracow, 'the very centre of Polish culture' into a newly created 'German' area was madness, and would only damage Germany's reputation.[4]

> The administration of western Galicia is overburdened with such problems and in chaos, while continuing uncertainty over the future nationality of this or that administrative area increases the existing chaos.[5]

The Police Chief at Bromberg, SS GF Hildebrandt, commented in November 1939 that room had to be found in his area for another 10,000 Wolhynien Germans, who were 'in the first place to be considered as successors to the Poniatowski villages',[6]—meaning presumably that the unfortunate Wolhyniens would be used as armed settlers to fight their Polish neighbours in the countryside. The argument in favour of using the *Volksdeutsche* as *Wehrbauern* tended to be produced, understandably perhaps, by SS Police Chiefs and by Himmler, while the argument in favour of peasant food production came from members of the National Food Estate drafted as agricultural advisers after the outbreak of war.[7] Members of the RuSHA advisory staff were strongest in their advocacy of deporting (or Germanising) the Polish population, and replacing them with Germans, for both racial and food production reasons. One suggestion that seems not to have been made was to try to increase food production using existing Polish agricultural organisational methods. Not only had numerous studies of Polish food production between the wars revealed the fact that production had dropped by between twenty and forty per cent in formerly Prussian areas,[8] but a special study commissioned by the National Statistics Office in July 1939 described Warsaw, Lodz, Silesia,

This tablet commemorating Darré, Blood and Soil, and Goslar 'the old imperial city', was first placed in one of Goslar's churches in 1934. (*Goslar Archive.*)

Goslar becomes Germany's National Peasant City. 1934.

Darré (second from left) waits to address a meeting in Goslar. Behind him the wheatsheaf symbol of the National Food Estate, superimposed on a swastika.

Above: This picture of Darré's birthplace in Belgrano, Argentina (1896), was sent to him in 1936 by a N.S.D.A.P. youth group in Argentina.
(*Goslar Archive.*)

Below: Darré here gives an after-lunch pep talk to the leaders of the National Food Estate, at a farm belonging to one of them. 1934/5.
(*Mrs. N. Backe.*)

Nürnburg postcard. 1935.

Top: Darré was shown the medieval ex-monastery behind the Imperial Palace at Goslar. Here he points it out to his wife. Darré was vehemently anti-Christian. (*Mrs. N. Backe.*)

Left: Commemoration stone at Goslar.

Right: Hitler addressed the Harvest Festival Rally at the Buekeberg. 1935. (*Mrs. N. Backe.*)

Troops parade through Goslar, the new peasant city. 1934. (*Goslar Archive.*)

Peasant labour used to build their own new houses, to cut costs in new settlements.

Galicia and Kielce as 'agricultural deficit areas'. Germany had double the number of cattle, and four times the number of pigs and goats per capita,[9] twenty-five per cent more arable land and twenty-five per cent less forest.

> More than one-third of the agricultural land belongs to farms of less than 2 ha; some one-fifth up to 5 ha. No surplus worth mentioning could be brought to market—even with sensible management—given these farm sizes. A large part of the remaining ground is forest. An example of what is produced by the larger farms: one 200 ha farm we took over, which on the whole had not suffered from the war, presumably thus not lost any poultry. Some 28 ha of fish pools delivered 8–10 cwt, of fish a year! The rye was enough for the agricultural labourers, and the manager grew his own food. You can see from this example what large and medium farms are going to produce for urban consumers.[10]

Another report to Himmler commented that Polish land was 'uneven, lacking in natural resources, exhausted and extremely wet, with no natural drainage'.[11] Thus, a mixture of contempt for Polish farmers for having fallen behind Prussian standards, and the fact that areas like Galicia were in any case among the poorest peasant farming areas in Europe,[12] led inexorably to the decision to deport Polish farmers and substitute 'good German peasants' in their place. A vivid impression of the poverty of the Polish countryside is given in the diary of a Nazi administrator in the incorporated areas in 1941:[13] 'Everything is primitive, poverty stricken and filthy. With the German settlers from Russia, things are much cleaner, but otherwise little different ... I was told that people actually lived in holes in the ground round here. I saw poverty stricken villages ... unplastered houses, reed thatched roofs, reed and clay walls ... farmhouses without solid floors or even plaster.'

In fact, the exent to which Poland was virtually an agriculturally undeveloped land was well recognised by local officials. However, the implications of this fact, the resources and effort that would be needed in order to make Poland and the incorporated areas net contributors to Germany's food requirements, were consistently ignored by administrators in Berlin, where eyes were fixed on the Polish rye exports in the last two decades. One report which showed the unfavourable conditions in Polish agriculture compared to Germany pondered on where the 'agricultural surplus area' was in Poland, concluding hopefully that it was in the west. But, 'the area [Warthegau] is in its current structure and density of population *a deficit area*, and in no way a corn chamber, the assumption which is heard all over the place in Berlin'.[14]

This report contained an early detailed plan to carry out the population transfer which was at once to turn the Incorporated Areas into an agricultural surplus area, and settle it with German peasants. He recommended evacuating double the number of Poles and Jews for the number of planned German immigrants; all Jews, and the Polish intelligentsia would be immediately deported, while the 'indigenous population' would be investigated for possible Germanisation. This meant that a large number of Polish citizens were envisaged as remaining in the area. Meyer, appointed Himmler's chief planning officer in late 1939, called for a population balance of fifty per cent Germans, and fifty per cent Poles, with a land holding of sixty to forty per cent.[15] Himmler, however, addressing SS leaders in Danzig in October 1939, prophesied that in fifty years' time, some twenty million German settlers would be living in Posen-West Prussia— although he wanted settlements to be kept at least 10 km from the Polish border.

The land would be free; cheap labour would be available from Polish workers. Forests would provide wood. All that had to be paid for was electricity, piped water, sinks and baths. Like most of Himmler's proposals for *Wehrbauern* settlement, this bears out Backe's comment that Himmler's settlement ideas were 'vague, theoretical and not based on practicalities'.[17]

The vaguer Himmler's promises, the more detailed were the recommendations of the RuSHA officials. Gustav Pancke recommended that peasant settlement be confined to western Galicia and the southern part of Upper Silesia, so that Polish industrial areas should be spared any evacuation measures. He requested the *Höhere SS und Polizei Führer* to direct all settlers to the south.[18] Hildebrandt also wanted to avoid disturbance to the Polish economy, and suggested drawing up lists of firms which were essential to the Polish economy, 'to avoid arbitrary disturbance'.[19]

As the months dragged on and homes had to be found for hundreds of thousands of *Volksdeutsche* refugees, the arguments over the best way to use the new 'colony' continued. One member of the SS planning department for land in the east visited Gauleiter Greiser in February 1941 with detailed maps of planned re-settlement in the

Warthegau. He argued that even if every centimetre of Polish land were to be farmed by Germans, the population would still be only one quarter German, because 'the removal of the Polish agricultural and urban proletariat will be a much more difficult task than that of peasant settlement',[20] and that unless the existing agricultural structure was changed, the existence of Polish farm labourers made 'a true Germanising' impossible. Gauleiter Greiser objected that 'the task of the Warthegau is to produce grain, grain and still more grain—a grain factory, and that is why it has been incorporated into the Reich', to which the RuSHA adviser replied with arguments so characteristic of the peasant producer-advocate that it is worth quoting in detail.

> I answered that fats and milk were needed more and peasant farms were more productive in these areas. While this was certainly no time to start experimenting, there was undoubted evidence that peasants produced more than large landowners from the same land, especially with the aid of machinery. Only the peasant could make the Warthegau a German county—he alone rendered the Polish labourers redundant.[21]

However, Greiser reiterated that Goering had expressly ordered grain production to be the main crop, using large farms and Polish labour. The aim of 'Germanisation' was excellent in itself, but would take thirty or forty years. With the inhumanity typical of the reformer mentality, the RSS adviser made the counter-proposal that Polish labourers' families should all be deported to the General Government in order to induce their menfolk to follow them. Finally, Greiser's 'reactionary adviser', Siegmund informed the Race and Settlement office that Goering had ordered that Poland's agricultural structure should remain unchanged, with the single exception of dwarf farm consolidation. Only the large estates could produce corn, or breed herds of pedigree cattle. An agricultural leading class was needed, and furthermore, the demands of wealthy men who wanted farms needed satisfying—something stressed by the Wehrmacht. 'It is mere fantasy to talk of 5,000 new farms by 1941. You yourself know how all these great plans of the RF [Himmler] end up'.[22]

This interchange of opinions tells us several significant things about the development of settlement plans in the incorporated areas. First, that even before the invasion of Russia, which put a stop to ambitious re-settlement projects, very little had been done in the way of Germanising the Wartheland as originally planned. Second, that

Goering's orders as head of the Four Year Plan, and minister in charge of the economy in the eastern territories, were to leave the existing Polish agricultural structure in place. Third, that Himmler's impressive sounding decrees and plans were widely regarded as being romantic fantasies, and should perhaps not be taken too seriously by historians today.

For example, in December 1940, one of Himmler's *Anordnungen* stated that

> to create opportunities for settlement, it is necessary in the Incorporated Areas to give peasant settlers a greater share than before of farms seized from Polish and Jewish hands—40 per cent instead of 25 per cent of the agriculturally usable acreage.[23]

whereas, the analysis of the implementation of these plans shows that Germanised land in the Warthegau comprised barely ten per cent of the agriculturally usable land by May 1941.[24]

Again and again, the need to ensure that the 'recaptured' province would become and remain German was stressed in writings of 1940–1.

> The character of this land is not determined by the formal property ownership of large estates and domains, as long as the necessary peasants and agricultural labourers are of Polish blood; a realisation achieved by Wilhelmine Germany far too late.[25]

wrote one agricultural expert, and 'Only the *Neubildung deutschen Bauerntums*' could 'neutralise the danger of having to use Polish elements'[26] as agricultural labourers.

One writer urged that all racial undesirables should be sorted out and put into labour battalions, but were on no account to be sent east, because Himmler planned a 'blonde province' there,[27] a remark that was also quoted by Pancke in a letter to Himmler.[28]

Between 1939 and 1940, therefore, the very period when the *Volksdeutsche* were due to be resettled, the exercise was still in the planning stage. Broadly speaking, the Race and Settlement's propaganda had a sharper racial content than that of writers like Kummer, Schöpke, Haushofer, and others who approached the resettlement from the small-peasant tradition of German *Agrarpolitik*. In 1940, the new head of the RuSHA, Hofmann, sent Himmler a copy of *Das Bevölkerungspolitische ABC*, which discussed the movement of populations in terms of the right to survive of the superior group,[29] whereas Kummer, in a series of speeches made in 1940, talked more of the value of German settlers to the nation's security, and quoted the old German folk-song that had become the theme song of the *Artamanen*—'Over the green, green heath, there is a better land'—as signifying the continuity of the *Volksdeutsche* re-settlement with the (alleged) peasant colonisation of East Prussia.[30]

Schöpke stressed the high fertility of the German colonies in Russia, with their population increase of ten to twenty times in a hundred years. He saw the ethnic Germans as a vigorous source of renewal for the agricultural population, whose presence would enable foreign agricultural workers to be repatriated after the war.[31]

Many of the writers who had propounded the virtures of peasant production supported 'Germanisation' of the Wartheland for the same reasons. Dr Otto Auhagen, whose dissertation in 1896 had compared the productivity of small and large estates, reported enthusiastically in 1940 on the possibilities of exchanging populations on Germany's eastern border. He referred to similar plans drawn up during the First World War.[32] Members of the Economic Geography Institute in Königsberg also drafted proposals for settling the 'New German Area' in September 1940, arguing that complete Germanisation could only take place through peasant holdings of fifteen to twenty-five hectares.[33]

Some saw the specifically productive aspects of the venture as a challenge. One writer described the flat, largely uncultivated plains of the Warthegau, exposed to cold drying winds, and lacking in hedges and shelter belts. The winters were long and cold, while roads, electricity, drainage and schools were largely lacking. But

> With the new land in the East, it will be easier to create new forms than in the *Altreich* ... a new beginning ... we do not have to struggle with established habits, as in the old villages of the West.[34]

Seldom can the modernisation drive that ran in tandem with the support for small farms have been so clearly expressed; significantly, this article appeared in *Neues Bauerntum*. However, the implications of what was involved in creating a viable, modernised agriculture in a conquered land seem not to have been considered to any extent; because of the emphasis traditionally and incorrectly placed on the shortage of land within Germany, the availability of 'free' land in

Poland was seen as a cure-all for all the shortages of resources that had beset German agriculture.

The next section of this chapter will show the extent to which the plans failed, although in terms of agricultural productivity, considerable improvements appeared under the leadership of RNS advisers. What is surprising in these early enthusiastic plans is to find Darréan ideas about the need for a Germanised medium-sized farm structure surviving in such a strong form at middle and lower levels of the SS, as if a form of counter-penetration had taken place. Himmler, with his plans to develop an industrial raw material empire in the east, had taken over re-settlement, and Darré had returned to obscurity. Yet many of Darré's supporters continued to work for the eventual triumph of their own agrarian aims in the incorporated areas, convinced that they had the support of one of the most powerful men in the Third Reich, secure in their dominant aims: to modernise agriculture by reforming its structure, and to reform the agricultural class system. Whatever Hitler's motives for the *Volksdeutsche* settlement, SS planning documents leave no doubt about what they thought was their function: to provide the necessary demographic material for Germanising the Warthegau; to strengthen the then border area; and to improve agricultural productivity by cutting up the latifundia.

The existing situation was a blueprint of everything the SS planners opposed, from the hierarchical class structure to the ungovernable and unproductive agricultural proletariat, the unused potential of the land and the racial/cultural mixture. In December 1940, a report of the result of one year's resettlement activity again attacked the idea of making the Warthegau a haven for a German upper-class landed gentry, using Polish labourers. This attack was from the SS Race and Settlement office, and demonstrates how deeply Darré's anti-Junker views had penetrated the new planning bodies under Himmler.[35] Indeed, it is possibly for this reason that the old RUS advisory staff had been dissolved by Himmler in late 1939.[36] Their criticisms of Goering repeated. almost word for word, remarks made by an RNS member at Darré's settlement conference in June 1940:[37] 'The faults of *Innere Kolonisation*, which created a German upper class using Polish labourers, to the detriment of the nation, must be avoided.'

These were Darré's points; however, unlike Darré's speeches, SS plans were couched in the language of practical politics. Phrases like *Blut und Boden* were missing, and the SS's opponents in this matter—Goering and the Wehrmacht High Commands—were seen as reactionaries with outdated unscientific beliefs, whose obsessions about the social and economic value of large estates would soon enter the dustbin of history. This belief in a more egalitarian farm and land ownership structure was seen as the modern, progressive and scientific norm. Clearly, this interpretation was worlds away from Darré's rationale for peasant farming, yet equally obviously it had been inspired and influenced by it. Himmler's romantic interest in medieval German history seems not to have linked up with the 'forward-looking' workers in the RuS and RKFDV; while in his practice, Himmler was unenthusiastic about altering the Polish farm structure

RESETTLING THE VOLKSDEUTSCHE

In September 1939, the ethnic Germans who were to be involved in the re-settlement programme fell, broadly speaking, into two categories. There were those ethnic Germans who lived well to the east of the incorporated areas, such as the Baltic Germans who had lost their land under the Estonian, Lithuanian and Latvian land reforms of the 1920s, and the more pressing problem of the German colonists from Bessarabia and Wolhynia, some 160,000 people who had originally been invited to settle by Russian and Russo-Polish nobles, after 1815 and 1863 respectively. Under the Ribbentrop-Molotov Pact, Russia agreed to repatriate these German colonists, who lay in the path of their invasion of eastern Poland.[38]

The second category was that of the ethnic Germans living *in situ* in Polish and Czech provinces which had been German/Austrian before the First World War, together with the 750,000 or so refugees who had left the Polish borderlands between 1919 and 1939, considered by Germany to be refugees with a right to compensation by the Poles.[39] The ethnic Germans still living in Poland were referred to as *Reichsdeutsche* by the German administration after their invasion of Poland, a fact that confused the issue at the Nuremberg trials, where the judges thought that the *Reichsdeutsche* in Poland were recent immigrants from Germany; the reverse was the case.[40] By 1942, it was estimated that some 500,000 people were involved in the resettlement programme.[41]

Despite Hitler's well-publicised remarks about the South Tyrolean Germans, for example, just sailing down the Danube to the Black Sea, and finding a home there,[42] the thrust of the re-settlement was defensive rather than expansionist, and was seen as a definitive consolidation of German territory. Homes had to be found for hundreds of thousands of refugees, and the incorporation of the newly conquered territory in the east was the answer to many problems. There was the hope that it might include part of the rye-growing area that had contributed to Poland's massive grain exports before the war; the population density was half that of Germany's, which meant, it was thought, free and empty territory for settling farmers,[43] while the *Volksdeutsche* would fill the gap caused by the lack of German émigrés from the *Altreich* prepared to go east. They would provide an indisputably Germanised population in land that would now be German forever. Never again could plebiscites and percentages be a weapon used against land colonised by German settlers.

This application of the concept of self-determination and ethnic identity as a basis for nationhood—a complete reversal of the expansionist liberalism of 1848, that proposed giving German nationality to all Germans anywhere on earth—was applied in a territory embittered by decades of strife. German claims of ill treatment of their minorities in the inter-war years were not a figment of Nazi imagination,[44] although, ironically enough, anti-German activities appeared to have receded in Poland after the German-Polish Friendship Treaty of 1934. There are great difficulties in judging the true extent of atrocities against German civilians after 1939. For example, the massacre of many Germans at Bromberg in September 1939 was inflated by Rosenberg to a figure of 50,000, while the post-war Polish government admitted to 300. Contemporary accounts, and the war-time trial of those responsible suggest that the real figure of deaths was over 5,000.[45] Militant Polish ex-soldiers and members of nationalist fighting units had been given priority as new settlers in the Polish border areas where the 'Poniatowski' villages were formed, a fact that makes the mutual hatred and fear between Germans and Poles on the border easier to understand.[46] These remarks are not intended in any way to excuse either the German invasion or the evacuation measures that followed it, but rather to help to explain the attitude of the German planners and administrators of the re-settlement programme, men who were not monsters, but who implemented a plan that caused endless suffering.

The total amount of territory annexed to form new German counties in the east was 102,800 sq. km,[47] of which some seventy-five per cent had been German territory before the First World War. German law was introduced gradually, including the *Reichsnährstand* Marketing Law, which came into effect in January 1940. Approximately eighty per cent of the twelve million population were

Polish, 4.5 per cent Jewish, and some 15.5 per cent German. The proportion of Jews within the population was considerably smaller than in eastern Poland, as the main areas of Jewish settlement lay in regions which had been Russian and Austrian prior to 1918.

The RKFDV decree of October 1939 authorised the Reichsführer SS to create new settlement areas, protect the *Deutsche Volksgemeinschaft* from 'damaging influences', and above all, organise the return of Germans living outside the boundaries of Germany.[48] Clause 3 of the decree gave the task of creating 'new peasants' to the Ministry of Agriculture, under the RKFDV's jurisdiction.

The RKFDV's creation was a shock to the Race and Settlement Office (now called the RuS) which had expected to run the re-settlement. Local staff offices had been established in Danzig, Silesia and West Prussia for the event, and detailed maps had been drawn up of each parish by the Race and Settlement advisory staff, work that was later taken over by the SS planning staff in Posen and Lodz under Konrad Meyer. Negotiations took place with the civil government in the occupied area in October, the office offering to place itself under the jurisdiction of the local governor, but Himmler set up a parallel organisation of planning experts which rendered the Settlement advisory staff effectively redundant.[49] In late November, 1939, they were transferred to the RKFDV under the aegis of the police chief in Danzig, SS Gruppenführer Hildebrandt, jointly to run Land Offices for the implementation of the 7th October 1939 decree in Silesia, Danzig and Posen.[50]

Pancke suggested immediately organising his now underemployed staff into *Vorkommandos* who would be sent into villages to evacuate the Polish inhabitants, and prevent them destroying livestock or crops. He accompanied this idea with a request for funds to cover weapons, transport, and so on, and an exemption from Wehrmacht military duty, and gave details of estates that could be taken over for use as SS training farms (*Lehrgüter*).[51]

This enthusiasm was restrained by a meeting with Himmler some days later, when the representative of the civil government, SS Freiherr von Holzschuher, was present. Himmler ordered that the proposed Land Office be subjected to the RKFDV office, now under SS Brigade-Führer Greifelt. As for the training farms, he agreed that suitable land (especially if formerly owned by the church) should be seized and prepared, but that no training of 'new peasants' should be carried out while the war was on. The plans went into absurd detail, but the memorandum of this meeting shows that while Himmler's directions about co-operation with the police, and his administrative orders, were clear, his ideas about the actual settlement of formerly Polish villages were not; and that Pancke's enthusiasm to begin operations was being heavily controlled from the centre.[52]

The Race and Settlement Office had been financed at the beginning of the war from the Prague Land Office, but by December 1939 this source of funds had dried up.[53] Pancke sent Himmler some of their reports, to 'show what excellent work they are doing',[54] settling South Tyroleans in Kreise Saybusch, but Himmler seemed, if anything, alarmed at the verve and speed of the operation: 'The work they have performed is certainly interesting, but to my mind is at this moment premature. Where exactly are these RuS *Beratung* people now, and what are they working on?'[55]

Later, he was to apologise indirectly to Pancke for the dissolution of the RuS advisory staff, and the takeover by the RKFDV:

> Your own position [as head of the RuS] does not play any special part in my work as *Reichskommissar*, because in fact I would otherwise like to give the RuSHA perhaps an even greater task; however, the joint role would endanger relationships between us and all the Ministries, which is not to be sneezed at.[56]

Between December 1939 and December 1940, events seemed continually to run ahead of planning, but, cumbersomely, the re-settlement proceeded, amid complaints that the SS was being starved of funds by the Ministry of Agriculture, and attacks on the Settlement societies for 'interfering with ethnic German re-settlement'.[57] Von dem Bach apologised to Pancke that only 1,000 people a day were being evacuated from Upper Silesia, instead of the planned 4,000. With no central water supply in most of the Polish towns, and the danger of typhus in the summer months, establishing transit camps was itself no easy task.[58] The planned rapid evacuation of Polish farmers was not taking place, and all large buildings that seemed suitable were bought by the RKFDV for temporary accommodation for the *Volksdeutsche*.

In December 1939, the plan was to evacuate 400,000 Poles to make way for around 200,000 *Volksdeutsche*, of whom some 120,000 were expected immediately, most on foot. Categories for deportation to the G-G included all Jews, Polish peasants whose land was being taken for the Wolhynien Germans, and any Poles related to those killed in the fighting, who were described as a security risk.[59]

The Wolhynien Germans did not in fact begin to arrive until the spring of 1940, after complex negotiations with the Russians, and the implementation of a refugee treaty concerning those Poles (some 60,000) who had fled to Russian-occupied territory after Germany's invasion, but who now wanted to return. RuS officials went to Russian-occupied territory to oversee the evacuation of the Wolhyniens and Bessarabians. Uncertainty surrounded their evacuation: Pancke claimed that he was not told to prepare for their arrival until February or March 1940.[60]

The colonists arrived with the minimum of luggage and possessions. They were confined to fifty kilos per family, and had concentrated on bringing with them seeds and farming implements. 'The negotiations with the Soviet Union went without friction; the officials were extremely *korrekt*', commented Hoffmeyer approvingly, although 'It was sometimes difficult to communicate with such completely opposed points of view. Also, one had to get used to the fact that time and punctuality in general meant nothing to the Russians.'[61] This meant hours or days waiting at stations at temperatures sometimes twenty degrees below freezing, while lost train carriages were found, and Soviet officials borrowed pencils and paper from the Germans. Even under war-time conditions, the incorporated area was a paradise of plenty compared to areas under Russian rule. However, more panic-stricken confusion awaited the unfortunate *Volksdeutsche* in Germany, where no one seemed to be in charge of the project, or to have any idea of how many people were involved.[62]

Pending the provision of suitable farms, returning ethnic Germans were placed in transit camps within the Reich as well as in the incorporated areas. The Waffen SS, with its chronic shortage of manpower, recruited as many of their youths as possible, while families were sometimes split up by recruitment by labour organisations to work in the *Altreich*. The National Socialist Welfare Organisation helped with the provision of furniture and clothes, while the RKFDV provided pocket money for those in camps, and other welfare services.[63]

Evacuation orders regarding Polish farmers were strict in their allowances of furniture and livestock; the farmers were supposed to leave behind heavy furniture and cows and horses. However, in practice, they seem to have taken all they could with them, while the buildings and crops that remained were often damaged by relatives of the evacuated Poles. In many cases, the evacuated Poles were moved only to the next village along, and the re-settled *Volksdeutsche* feared their presence.[64]

The returnees varied considerably in their experience of farming, their abilities and education. Bessarabian Germans turned out to be especially difficult to settle in farms, and in December 1940 Himmler ordered them to be put into labour battalions instead.[65] Relations between different groups were sometimes tense, as indeed they were between returnees and the old-established *Reichsdeutsche* families in the incorporated areas.

Himmler had ordered in January 1940 that 'despite the joy these Germans feel on returning to their homeland, every encouragement must be given in order to make their adaptation to the new surroundings, and the rebuilding of their lives easier'.[66] But among the substantial element of integrated *Reichsdeutsche*, usually estate owners and managers who had intermarried with Poles over generations, the newcomers were despised rather than welcomed as racial comrades. In the eyes of the SS administration, many of these integrated Germans had 'gone native', and become quite Polish in their lifestyle; they presented a special ideological problem to the administrators of the newly conquered territory. 'Those *Reichsdeutsche* who own or manage large farms have personal friendship with Poles and drink with them. Unfriendliness towards the *Volksdeutsche* is a commonplace.'[67] What was even worse, some Poles were too friendly with the *Reichsdeutsche*. Local NSDAP welfare workers found their sense of nationalist propriety offended when they heard, for example, Polish maids singing German nationalist songs in a *Reichsdeutsche* household.[68] Class loyalty appeared to be a stronger force than racial solidarity in relations between the new immigrants and the old, at least, where large and medium landowners were concerned.

By mid-1940, thirty-five camps had been established around Lodz and other towns in the eastern zone of the incorporated areas, most of them adapted from empty factories and summer homes. 120,000 people had passed through them by summer, while by the end of 1940 some half a million ethnic Germans altogether had 'heard the call of the Führer and returned to the Greater German homeland'.[69] Less than half, however, had found farms, possibly fewer, if the figure given was inflated for publication. Nonetheless, as the report pointed out, that meant that an area of land the size of Oldenburg had been 'Germanised' in West Prussia, Upper Silesia and the Wartheland. Up to this date, the idea of a peasant settlement was still dominant. 'The work has just begun. The external incorporation of land must march together with internal Germanisation ... *Volk* borders are more decisive than national ones'.[70] Some ethnic

Germans, however, were reluctant to return to rural life, especially the Baltic Germans.[71]

The shortage of labour in agriculture and industry—in Germany especially—overtook re-settlement plans by the end of 1941. The transports of Poles to the G-G were suspended in March 1941 because of the shortage of labour[72] by which time some 400,000 Poles had been evacuated and 167,450 ethnic Germans settled in their place. The SS Planning Staff at Posen called in vain for further evacuations, arguing that after the war was over, American-European relations would be at a low ebb, and that refugee German settlers from North America could be expected to arrive![73]

The Planning Staff did their best to help the *Volksdeutsche* settle in, providing libraries and language courses for them, and attempting to fit suitable tradesmen to the right jobs.[74] The attitudes of the Poles varied from entirely hostile to passive, with even some active friendliness shown to the new arrivals in their often desperate straits. Greifelt commented, 'In general, the Polish section of the population have not shown themselves to be friendly to the German resettlers and the resettlement commandos, but have also not been directly hostile.'[75] This situation was not to last.

The easiest answer to the joint problems of shortage of labour and Polish returnees (both from the Russian-occupied area and the G-G), was to conscript Polish men and women as forced labour, and deport them to the West, to work in Germany and France.[76] Polish dwarf farms provided most of the workers, and the smallholdings that were vacated were then consolidated for German resettlement. In fifteen parishes around Lodz, over 500,000 hectares were resettled in this way (half by *Volksdeutsche* peasants, and half by returning Germans from the old Reich). The SS planning staff at Lodz complained that this still left 750,000 hectares locally in the hands of Polish smallholders, which could be made into 40,000 medium-sized farms, which 'would break the Polish influence', and enable the land to be used for intensive fodder production, more animal rearing, sugar beet and vegetables.[77]

Obviously, RNS ideology had found fertile ground in the SS Planning Offices at Posen and Lodz, and indeed there was a considerable increase in food production in the occupied areas under their management; however, whether this was due to resettlement or to the doubling of tractor numbers and import of ploughs and drills from Germany, is unclear.[78]

The implementation of more intensive and modernised farming encountered the problem of land shortage. When the Germans invaded the Wartheland, there were 3.2 million agriculturally usable hectares, but the first priority in a plan drawn up by the RKFDV Land Planning Office was to set aside nearly 500,000 hectares for afforestation, as well as 100,000 hectares for the Wehrmacht, and 620,000 hectares for new roads, industrial areas, suburbs and so on. This immediately cut down the available agriculturally usable land to 2.6 million hectares. 800,000 hectares of this was allocated in 1941 to the *Volksdeutsche*, although they settled only some 300,000 hectares.[79]

Despite the available free land, the existing houses and buildings which had belonged to evacuated Polish farmers, attempts at intensifying farming suffered from the lack of labour and machinery that had beset German agriculture in the *Altreich*. Difficult climatic conditions were exacerbated by the lack of hedges and copses; a long winter and a dry spring produced problems that had to be solved by altered farming methods. 'The aim must be to confine cattle rearing to one cow per two hectares, and have an intensive arable rotation of sugar beet, catch crops and potatoes for pig feed', wrote Hermann Priebe in *Neues Bauerntum*. He attacked ideas about peasant autarky as mere sentiment, and called for more machinery, artificial fertiliser and aids for milk production to be made available to the new farms of the Warthegau.[80]

Under the trustee system, whereby German farmers were sent out to manage Polish farms, production improvements did come about. However, by December 1942, re-settlement had virtually ceased (which did not stop Himmler laying down directives for cultivation and planning).[81] By December 1942, only the 35,000 or so new *Volksdeutsche* farms remained of the 'blonde province'. Every aspect of the original plan, save only the increase in food production, had failed. Even racially, 'in exceptional circumstances, persons of mixed race are to be admitted to the Reich'.[82] Germanisation in the incorporated areas meant in effect the attempted Germanisation of some eighty-five per cent of the existing population there, while those *Volksdeutsche* still in transit camps at the end of 1942 were utilised for labour battalions in Germany.[83] Perhaps fortunately, they had developed 'a certain fatalism as a result of war and revolution'.[84]

This necessarily brief look at the *Volksdeutsche* re-settlement programme has obviously ignored many factors which are important in a wider context. One problem is that the welfare aspect of the programme was considerable, and dominates the archival material. As Koehl points out, the picture that emerges from the RKFDV and RuSHA documents is overwhelmingly one of humaneness, with caring personnel distributing pocket money and showing ethnic German wives how to mend clothes. This took place, however, against a background of forced evacuation, dispossession, draconian anti–Polish legislation and racial bitterness that must be mentioned here, if only to clarify the position of the new settlers who are the subject of this section.[85]

While frequently the subject of personal kindness from the Polish majority, the ethnic Germans were also obliged to build up homes and farms from scratch, in conditions of poverty and shortage. Many no longer spoke fluent German, but were expected to demonstrate the virtues of German culture to their Polish neighbours. Understandably, they were subject to sporadic physical attacks from partisans and Polish villagers. Far from being the new German homeland which they had been promised, the incorporated area was an alien territory of flat wastelands, a wilderness stretching endlessly into mournful sunsets, where the unprotected emptiness of the eastern horizon exacerbated the fundamental agoraphobia of many of the settlers.[86]

It emerges clearly from the reports, diaries and memoirs of the German administrators that the resettled Germans suffered a poverty as great as that of the Polish villagers, despite the efforts of welfare visitors and other officials. Furthermore, many of the new settlers came from farming communities where techniques were primitive. New methods had to be learnt, with concomitant delays in successful production.[87]

For most of these groups, their eventual fate was worse than if they had remained behind, for the end of the re-settlement period was poignantly tragic. Driven west by the advancing Russian army, murdered *en masse* by Polish partisans, subject to every untrammelled atrocity that troops fuelled by years of hate propaganda and reaction to the terror of the *Einsatzgruppen*, could devise, the degree of their destruction will probably never be known for certain.[88] Along with the many millions of East Prussians who fled west in 1945, all that was left of the ethnic German returnees at the end of the war was an incoherent, fragmented body of survivors. Not only did several hundred years of European history disappear with their physical demise, but their fate was to be either ignored or interpreted as the just deserts of German imperialism.[89]

German agricultural settlement was not an integral part of Himmler's institutional expansion, but was an obsession of his staff, and popular with the majority of SS members. Himmler appears to have seized the opportunities offered in 1938 when settlement ideology was conjoined with the sudden prospect of territorial expansion. Jan Gross suggests in his work on Poland under the Nazis that the Nazi creed was incapable of coping with the concept of empire, that both Nazi racialism and its emphasis on institutionalised 'personalism' was incompatible with the creation of a real 'New Order'.[90] Himmler's importance lay in the fact that he, almost alone among many other Nazis, was ideologically and temperamentally capable of thinking in imperial terms. This helps to account for his rapid seizure of power in the occupied territories; Himmler was a new phenomenon in National Socialism, representing its (perhaps inevitable) transformation into the full Fascist state: imperialist thus anti-nationalist; élitist not populist; seeking the efficient, planned—and rootless—European super-state. This dimension helps to account for the fact that by 1944 more than half the Waffen SS were non-German. The role of the *Volksdeutsche* settlement in Himmler's plans was to provide a valuable element of apparent continuity of aim through Konrad Meyer's Land Office; through the appeal to the old German desire to recover its lost emigres and consolidate its frontiers. It helped to camouflage the qualitative distinction between SS hegemony in occupied Europe and previous ambitions,[91] but its real purpose was to open the door to strategic control of a new empire in the east.

In the pursuit of that empire, Himmler discovered that racial purity could give way to a supra-racial and supra-national categorisation that magically enabled a vast source of manpower to become available. By re-labelling, people could be drawn into the system and ranked on a scale of Germanism. The concept lost racial and national meaning, and became a means of grading usable human material, 'began to acquire an achievement dimension'.[92] The *Volksliste* became a sifting procedure to procure potential citizens of the New Order: loyal, healthy, and possessed of five fingers on each hand.[93]

For many SS personnel in Himmler's empire, the well-being of the ethnic Germans had been of intrinsic importance; the correspondence between SS offices stressing importance of settlement as late as January 1945 is evidence to that effect.[94] However, control of the ethnic German re-settlement programme by Himmler played another role in the SS's drive to total dominance; by enabling them to control land use and distribution in the occupied territories, it put them in control of possibly the most vital raw material in all the conquered territories—land itself. There was no inevitability about the emergence of this domination. Himmler had to compete against Rosenberg, Commissar for the East, the Four Year Plan civil authority, the

Wehrmacht High Command administration, as well as the Foreign Ministry and Ministry of Agriculture. The story of that struggle is not part of this work; what is involved here is the interplay of agrarian and racial ideology behind the most comprehensive attempt to institute German peasant settlement between 1933 and 1945.[95]

That barely a third of the returnees received farms, and that the remainder worked in German factories or lived in transit camps all over the *Altreich* shows, once again, the extent to which the apparent ideological aims of the Third Reich were subordinated to the aim of maintaining control, of survival at all costs.[96] There can be no doubt that the re-settlement of some 250,000 people within eighteen months was a major achievement compared with the lethargic New Peasant settlement over the previous seven years; yet after the invasion of Russia, when the war can be said to have entered its most ideological stage, re-settlement virtually ceased. When the aim of racial and national rescue competed with the aim of imperial expansion, the latter came first.

Clearly, there was an ideological component in the re-settlement programme, which affected its mechanics in particular. It was assumed that peasants would continue to farm, that a peasant territory would be more truly 'German' in a populist sense. It was also an improvised response to a sudden emergency, a 'dictated option'[97], and the fusion of Darréan *Blut und Boden* ideology, demographic-national aims, and agrarian reform in coping with this emergency has been the subject of this chapter.

In 1940, Himmler had announced that German (*Altreich* and *Reichsdeutsche*) settlement would have to wait until after the war.[98] Barely one year later, the invasion of Russia completely terminated any possibility of implementing the *Volksdeutsche* programme,[99] so that the revamped version of agricultural settlement suffered the same fate as the earlier programme. Nevertheless, it has been suggested that German agricultural settlement was a cause of German expansion into the Ukraine and Russia.[100] While food production certainly was important in Hitler's calculations, German settlement there seems not to have been envisaged.[101] The bulk of the re-settlement planning was carried out—often after the event—as has been described—between 1939 and 1941; it faltered after that date, and settlement ideology can hardly be blamed for Himmler's exploitative aims in the east for a new 'source of raw materials'. As for the deportation of half a million Poles from the Warthegau, the reduction to ghettos and final evacuation of its Jewish population—the first was only in part connected with the planned exchange of population for the purposes of the re-settlement, the second not at all. The conscription of Poles for labour in Germany was more a consequence of the untrammelled, exploitative ideology of the Nazis at war than of any rural ideology, and certainly not Darré's.

Acts of revenge taken on Poles who had been members of nationalist groups; the oppression and corruption of a conquered people after an invasion: these also can hardly be laid at the door of the plan to re-settle ethnic Germans in the incorporated areas. Yet because of the plan's emphasis on race, as opposed to nationality, it has attracted especial vilification. Clearly, any programme of forced evacuation and dispossession involves great suffering. But whether the programme was inspired by a desire for national expansion, ideological hegemony or racial consolidation (as exemplified respectively by Polish, Russian and German expansion between 1918 and 1945) hardly affects the experience of the event either for those who suffered or for those who survive.

CHAPTER EIGHT

The Green Nazis

Today it would be difficult to ignore fears about erosion, the destruction of animal species, anxieties about factory farming, the social effects of technology and the loss of farmland. Such issues are discussed constantly in the mass media, as well as in the output of special interest groups. When Rachel Carson wrote *Silent Spring*, she focussed the world's attention on the ecological destruction caused by pesticides and other chemicals in the lakes and earth of North America. What has become known as the 'ecological movement', especially the party-political 'Green' aspect, implies also a broadly-based cultural criticism of the development of western civilisation. Its emphasis on technology, foreign trade, the division of labour, its urbanisation and inherent anomie are all seen as social and spiritual evils. Labour-intensive energy conservation, autarky, re-cycling resources and living close to the land are seen as inherently good.[1]

On the whole, ecologists do not call for a return to pre-industrial ways of life as such. They tend rather to stress research into new forms of technology which are more suitable to small communities, and which would avoid damaging the balance of nature to the extent observable today. Where ecologists have moved into party politics, they are associated with international and pacifist sentiment which is collectivist in spirit, and propounds re-distribution of the earth's resources—sometimes re-distribution of life's miseries—with the zeal of an inquisitor. These ideas are more dominant in northern Europe and America than elsewhere. Although there is a small ecological movement in France, it has remained inconsiderable in terms of its political and social influence. It is in Germany, above all, that the Greens have obtained their greatest influence and publicity to date.

It is not widely known that similar ecological ideas were being put forward by Darré in National Socialist Germany, often using the same phrases and arguments as are used today. He began to campaign for these ideas, especially organic farming, from 1934 onwards, and during the Second World War stepped up the effort to introduce organic farming methods into Germany. After the war, as a broken, discredited politician, he continued to write about soil erosion, the dangers of artificial fertilisers and the need to maintain the 'biomass', until his death in 1953. Two decades later, these ideas about man's relationship with nature and the organic cycle of animal-soil-food-man known as organic farming, had gained wider attention. Today, they have crossed party political boundaries, and adherents can be found across the political spectrum. They have, in fact, become part of everyone's mental furniture. Were Darré's ecological ideas integral to his other views, or were they irrelevant? Was it just an embarrassing accident that he should have hit on questions which preoccupy us today? Was there a serious ecological movement in the 1930s that was able to co-exist with National Socialism? Or was it in a sense 'alternative', as—I have argued—could be said of Darré's ideas?

The two decades before the First World War saw the growth of the *Wandervögel*, the roaming bands of students who took to the woods and mountains of Germany in search of new ways of life. They were opposed to urban anomie and alienation. At the same time, the works of Rudolf Steiner and the Anthroposophy movement were gathering support. While accepting Darwinian evolution, Steiner, an Austrian Catholic, propounded spiritual, vitalist ideas, and feared the despoliation of the earth. His ideas included astrology, reincarnation and the importance of magnetism. But it was his opposition to the exploitation of the soil that was the most influential element. The German Youth Movement had a potent back-to-the-land element, that was more practical and communally-oriented than Steiner's individualistic emphasis. After the First World War, several agricultural settlement groups were formed. They had a Tolstoyan flavour, and one quoted Gandhi's attack on industrial society, 'Machinery is the Greatest Sin'. Self-help and the Spartan ideal characterised these groups, which aimed to resettle the German borderlands in the east. Their magazines carried songs and pictures in the spirit of the pre-war Youth Movement, whose illustrator, Fidus, lived in a wood with a large family, rather like Augustus John. Fidus, described by one writer as a '*Jugendstil* hippy', was an old man when the Nazis came to power: Darré sent him enthusiastic greetings in 1938 on his seventieth birthday.[2]

Much of Darré's influence among serious agriculturalists came from the fact that his programme combined economic as well as moral arguments in favour of the peasantry. However, his attitude to technology as it affected the peasantry (technology which was needed for higher productivity) varied. For example, in a post-war work,

Darré, writing under the pseudonym Carl Carlsson, attacked technology for dissolving ancient forms of being. It created new forms of social organisation—the market as overlord, the division of labour (on and between different farms) and an exploitative attitude to the land. He described the petrol-driven engine as a technological form that should have profoundly altered the relationship of peasant to technology: it did not do so because industry and government were obsessed by large farms, huge machines and the examples of American and Russian large-scale farming.[3] Here again we come to a further facet of Darré's opposition to the entire structure of German life as it had developed. Monolithic size, latifundian scale—the whole concept of Spengler's megalopolis, were in Darré's view the result of culturally destructive foreign influences: and all institutions—law, church, education—were the product of the victors, and must go. This was clearly an anti-conservative stance, and the similarity of these ideas with today's Third World anti-colonial movements is not surprising. Germany in the 1920s was reeling from defeat and a cultural inferiority complex. This helps to explain the explosion of dissatisfaction, radical revolt, and desire for a new society that characterised Weimar Germany.

Steiner's ideas did not put the peasantry first, nor did they show much interest in farmers as a group. His emphasis on individual personality and development, astrology and reincarnation, made no appeal to Darré. In fact, he had a running battle with Seifert, a landscape architect who specialised in 'embedding motorways in the landscape ... organically', and who worked for Dr Fritz Todt. Todt accused Seifert of being a fanatical ecologist in 1936, but by 1939 the two men had compromised.[4] Seifert was probably the unnamed but influential ecologist in the Todt Organisation who, according to Backe, in 1939 persuaded Hitler to put a stop to any further land improvement in Germany, on the grounds that drainage and similar projects would ruin Germany's water table (shades of the Amazon Forest today).[5] Seifert criticised Darré's land improvement schemes, arguing that they would lead to a dust bowl, as in North America. He described the modernisation of German agriculture that was taking place under Darré as a disastrous interference with ecological balance, which would lead to fungal and insect infestation. He wanted agricultural practice to mimic nature according to the most stringent organic farming prescriptions, avoiding weeding, ploughing or monoculture. Wild plant life should be left to provide a source of disease resistance potential.[6] Darré responded that the RMEL was concerned for Germany's ecology 'as far as this is possible, without endangering the safety of the nation', and described Seifert's articles on magnetism and other Steiner ideas as 'false, fantastic scribbling'. Seifert's view on the benefits of wild nature was a 'Rousseauvian concept', then, as now, an insult among nationalists, implying a Utopian and simple-minded optimism about Nature and the natural man. 'Healthy' and 'sick' were man-made concepts, he argued; insects and bacteria were as 'natural' as their prey, but man had the right to protect himself.

> While the battle for life among all species expresses itself everywhere freely in the free state of nature, this is not permissible with cultivated plants, because man must intervene for reasons of self-preservation in order to guarantee ... its harvest for man's own food supply ... The present want of space within German territory forces us in ever increasing ways to make the soil more and more serviceable.[7]

Here, Darré disassociated himself from the 'Nature-before-man' arguments of the extreme ecologists. His emphasis on the essential artificiality of human life, the need to direct nature, parallels his belief in the need to protect man, the domesticated animal, from the effects of domestication. This, as Konrad Lorenz's biographer has argued, was a common view at the time.[8] Indeed, it is hard to dispute the fact that man no longer exists in a state of nature. This was one of the preconceptions of all interventionist social reformers; that, once away from a state of nature, man needed scientific guidance and control. Darré's twist to the argument was that it was necessary to live more naturally, to be closer to nature, in order to guide and control the harmful effects of over-civilisation. In his enthusiasm for using the latest scientific discoveries to improve society and become less scientific, Darré showed his inheritance from the Darwinian radicals of late nineteenth century Germany, and especially Haeckel, founder of the Monist League, and much admired by Bölsche, the populariser of biology. Haeckel's views on the nervous system fused Darwinian natural selection with some Lamarckian factors, and towards the end of the century, his monist materialism became more and more a monist vitalism: his holistic view of the world was similar to Darré's (indeed, he is supposed to have invented the word 'Ecology').[9]

Steiner was an enthusiastic supporter of evolutionary theory, and wrote several essays on Darwin and evolution before the First World War. He felt that evolutionary theory could blend with spiritualist determinants. Steiner's followers were to branch out in several directions, including Theosophy, and the religious character of these developments irritated Darré.

However, he gradually realised that many of Steiner's arguments could be useful, and he became prepared to adapt them, announcing that he would rename 'Biological-dynamic agriculture', the Steiner term, 'organic farming'.[10] He commented in August 1940, just after his organic farming campaign began, that he saw the peasant as 'a biological function in the body politic (*Volkskörper*) ... The peasant's agricultural activity [is] not just a matter of production, but a means of maintaining the Idea of Peasantness itself.' 'Biological-dynamism' was compatible with peasant farming. In its complete disavowal of industrial projects—artificial fertiliser, mass-produced grain, insecticides—it rejected industrial capitalism. The alleged discovery of a letter from a major chemical firm, proving that it had instigated an attack on organic farming, helped to confirm Darré's stance (unfortunately, the letter itself disappeared from the files after their capture by the Allies).[11]

Whatever the emphasis, Darré always rationalised his views. He claimed in public to support organic farming because it seemed the sensible sane way to farm, producing nutritious food without damaging the soil; privately, because it helped the peasant cause. He supported methane gas plants and suitable machinary for small farms. In 1935, he claimed that 'our forefathers had always unconsciously venerated trees and other living plants. We now know that plants gave out useful and desirable chemicals, so that old plant physiology was unconsciously very efficient.' This remark was followed by two pages of chemical analysis of plants, drawn from a technical article in a scientific journal; a good example of Darré's tendency to justify ideas with scientific data, presented in a journalistic but well-organised way.[12]

Steiner's philosophy attracted several prominent NSDAP members, including Rudolf Hess, and lesser functionaries, such as Dr A. Ludovici, land planning officer in Hess's office, and Ludolf Haase, brought into the Ministry of Agriculture after 1942 by Backe.[13] Steiner's mysticism, individualism and dislike of organised politics made his followers suspect to the Gestapo. The movement had Utopian and pacifist tendencies, and formed a rival neo-religion of powerful persuasiveness. Heydrich criticised it in October 1941, in a letter to Darré, for being an élitist philosophy: 'Not a *Weltanschauung* suitable for the whole nation, but a special teaching for a closely confined circle', and remarked that although currently masquerading as a nationalistic German ideology, it was fundamentally oriental in nature, and unsuited for 'a Germanic people'.[14] Backe's papers contain a cryptic message from Gauleiter Eggeling, the man in the powerful position of Keeper of the Seal after 1937, referring to Himmler's interest in 'organic ideas' after Darré's fall.[15]

Darré wrote several articles on ecological themes between 1933 and 1936. Some were on the dangers of erosion, some on the lessons of the depression for agriculture. His work looked at the example of America, and dwelt on the anti-erosion measures carried out under Roosevelt. Large farms, he argued, had been hit harder than smaller, more self-sufficient farms. It now might be more 'economic' to use horse-drawn ploughs than combine harvesters.[16]

As the Nazi government began to tighten up on its opponents many of Darré's supporters began to face harassment. Darré, by now, faced suspicion. But contemporary accounts refer to him as a prominent and popular figure. While Darré fought in meetings for higher food prices and more resources for agriculture, farmers who followed Steiner's ideas were pestered by local NSDAP men.

Food production and self-sufficiency were a major preoccupation for Germany at this time, and the *Reichnährstand* was interested in ways of increasing both. Although not a follower of Steiner himself, Darré had several supporters of Steiner on his staff, and was exposed to their arguments. While Steiner's emphasis on astrology, reincarnation and magnetism made no appeal, Darré realised that it was opposed to 'liberal, mechanistic' ideas. He lifted from the movement its belief that the soil was a living organism, part of a vital cycle of growth and decay. If this cycle was tempered with, valuable nutrients would be lost, and affect the food eaten by man.[17]

Steiner asked Dr Eduard Bartsch to start a farm on organic lines, in the sandy soil of the Mark. Bartsch founded a society for organic farming, and was also an active Anthroposophist.[18] The experimental farm at Marienhöhe kept crop returns and studied farm figures in their monthly journal *Demeter*, together with appeals for more hedgerows, shelter belts, drainage and organic compost. Occasionally, the inner spirit of the movement was revealed by quotations from Steiner:

We need a better knowledge of man,
Health through natural living,
Harmony between Blood, Soil and Cosmos,
Life reform as a national aim,

Knowledge and Life,
The Rule of the Living.[19]

Undeterred by party hostility, the organic farmers campaigned to win over Darré and his staff. They invited Nazi ministers to visit their farms. The National Food Estate held a meeting in 1937 with Hess, Bormann and Darré, to discuss the question.[20] At this time, the 'Battle for Production' for more food was in full swing. It emphasised increased use of artificial fertilisers, together with better seeds and livestock. Hess, who, together with several members of his staff, supported Steiner, was in favour of experimenting with organic farming, while Bormann strongly opposed it. The meeting ended inconclusively, but by 1940 many RNS members had been converted by the results at Marienhöhe, particularly after its humus–rich soil managed to produce a good crop during a prolonged drought. Reports favouring organic farms appeared in *Landpost* in 1940.[21]

Darré seems to have become fully converted to organic farming by May 1940. He realised that organic farms were more self-sufficient in terms of bought–in fertilisers and insecticides, and hence more in line with the self-sufficiency aims of the Ministry of Agriculture. The anti-capitalist implications appealed, too. Organic farming rejected industrial capitalism as well as the products of the big chemical companies. Darré organised a campaign to persuade the top Nazi leaders to support 'biological-dynamic' farming. He argued, ironically in the circumstances, that 'now that the war is over, we can concentrate on these matters', a phrase which gives some idea of his lack of contact with the top Nazi leadership at this stage. He circulated a questionnaire among all Gauleiters and Reichsleiters, disassociating himself from Steiner's mystical ideas, but proposing that information should be collected on the virtues of organic farming. He had prepared his campaign carefully, squeezing a loosely worded promise out of Goering that he could go ahead without interference, and stressing that he fully supported the principles of the 'Battle for Production'.[22]

Surprisingly, given the tense moment in Germany's history, the campaign aroused a great deal of interest. All the leaders replied except Hitler. Out of the replies still on file, seven were in favour of organic farming and Steiner, three unsure, three hostile because of the link with Steiner, and nine just hostile, mostly because of the wartime conditions. Surprising support emerged in the form of Hjalmar Schacht, Darré's old enemy. He told Darré that he was in complete sympathy with his views on German agriculture, despite their disagreements 'on details' in the past: and was horrified at the 'hypertrophy of the agricultural sector', the 'capitalistic concentration in individual hands' of recent years. He regretted the extent to which Darré's originally good concept had been turned into its complete opposite, and offered any help he could. A strange irony here; no less an opponent than Hjalmar Schacht, apparently won over to Darré's viewpoint, at a time when it was too late for either man, and far too late for Darré's *Agrarpolitik*. Bormann and Goering both reacted angrily against what they saw as an attempt to interfere with existing methods of food production.[23] Darré tried to see Hitler, but was warned off by Backe, who advised Darré that Hitler had ordered protection to be removed from ex-members of the Anthroposophist Association.[24]

The campaign was interrupted by Hess's flight to Britain in May 1941. On 9th June, the Gestapo seized anyone who was known to support Steiner, including Hans Merkel of the NFE, later Darré's defence lawyer at Nuremberg, and Dr Eduard Bartsch, editor of *Demeter*, copies of which were seized and destroyed. The few remaining copies today are in Himmler's files. Pharmaceutical works making organic products were also closed, with the exception of the Weleda factory (still in existence) which Otto Ohlendorf managed to keep open. Seifert complained to Darré that all members of the National Union for Biological-Dynamic farming had been searched. Despite a letter from Bormann, warning Darré that Hitler was behind the arrests, Darré, undeterred, appointed a working committee on organic farming, and wrote to Himmler and Heydrich asking them to stop arresting and harassing his farmers (and others, such as the nudists).[25]

Some extent of Heydrich's view of organic farming allied to the panic of his post-Hess investigations can be gauged from this extract from a letter to Darré of 18th October, 1941:

> I have had far-reaching investigations made into the connections between Anthroposophy and biological-dynamic farming in recent months. From a report on Anthroposophy, which I enclose, it appears that bio-dynamic farming emerged from the spirit of Anthroposophy, and can only be understood in connection with it. In practical terms it is obvious that Dr Bartsch, especially, of Bad Saarow, is at the same time a dedicated follower of Anthroposophy, and the leading representative of bio-dynamic farming ... Despite its temporary appearance of German nationalism, Anthroposophy is essentially oriental in its nature and origin.[26]

Even Darré's ardent supporters felt that his timing was faulty and that he had gone too far. One member of his staff described his action as 'grotesque and politically dangerous'.[27] But after the flurry of arrests caused by Hess's flight had died down, Darré continued to collect material on 'biological–dynamic' farming.

Bearing in mind Heydrich's persecution of the Anthroposophists and all their works, it is surprising that so many did support organic farming. At this time, the Anthroposophists were the dominant proponents of these ideas, and it is interesting to note that the brave handful of top Nazis who resisted Heydrich in this matter had their children educated and cared for by Anthroposophists after the Second World War.[28]

The campaign for organic farming was too much for Bormann to swallow. He had been investigating the Steiner groups before Hess's flight to Scotland. By March 1942, he had persuaded Hitler to demote Darré formally as Minister in favour of Herbert Backe.[29] It is surprising that Darré had lasted as long as he had, and this was partly due to Hitler's notorious reluctance to be seen to discard Ministers. Darré regarded Hitler's conduct of the war as bound to lead to defeat, and said so. He continued to complain that Hitler had 'betrayed Blood and Soil ideas'. His dream of a Jeffersonian republic of small farmers seemed to belong to another era.

One keen organic farmer, Rolf Gardiner, had visited Darré in Goslar before the war. He owned an estate in Dorset which he wanted to make into a centre for rural revival. In 1940, he broadcast on the BBC, describing his previous admiration for Darré and his ideas, and attacking the use made of them by the NSDAP.[30] He contacted Darré in 1951, and told him that the development of the English organic farming movement had been due to his inspiration. According to Gardiner, a group of like-minded people had met at the beginning of the war to found the 'Kinship for Husbandry' movement which had a cross-membership with the 'Soil Association', started at about the same time, and the most influential organic farming group in Britain.[31] Others have referred to the same group and its meetings in 1939–40, and several members seem to have been pro-Darré, although not pro-Nazi; so this statement may well be true. The English sympathisers also had experimental organic farms. Gardiner, incidentally, tried to start a 'Rural University' on his estate, which later became a trust for organic farming and reafforestation. John Stewart Collis, author of *The Worm Forgives the Plough*, worked there during the Second World War.[32]

Demeter was revived after the war, despite the disappearance of

Marienhöhe into Russian-held territory. Today it is a symbol of quality for health-food products, and the movement has branches in America and Australia. Sometime after the East German revolt of 1951, Bartsch's son wrote to Darré that 'the spirit of Marenhöhe still lives and breathes'.[33] These words, perhaps intended as comfort, were prophetic.

CHAPTER NINE

Catharsis

It is some indication of Darré's continuing importance in the general public eye that his dismissal of 1942 was kept as secret as possible. A brief statement was issued explaining that he had temporarily left office due to illness. Backe refused to accept the title of minister until 1944, and retained his title as leader of the ministry. 'He is the most suitable person, despite all his faults of character,' admitted Darré. Interestingly, the dismissal was mentioned by the BBC German language service, who broadcast a version of events highly favourable to Backe, and stressing Darré's reputation as a useless dreamer, and Backe's as a practical man. It seems obvious on reading it that detailed gossip about events in the Berlin ministries was available to the writers.[1]

The excuse of illness was widely accepted. Darré's health had in fact been poor since 1937. He suffered from asthma, eczema, the liver trouble that would eventually kill him, and all the ills of the chronically politically frustrated. Already by 1939 he had found himself alienated from the world around him, even to the extent of shrinking from entering a bar or café. Eyewitnesses describe how where Goering's appearances at Goslar would attract warm applause, and afterwards he would sit drinking with his supporters, Darré would create a circle of withdrawal, so that the party spirit would appear only after his departure. He was by now a hyper-sensitive, shrinking man, maintaining his self-image only through sporadic fits of rage and aggression, and even, quite unrealistically, contemplating joining up in 1940, to 'escape mechanical coercion, the appointment book, everything that ties me to the office by the ankle', to 'recreate the spiritual Western Front against capitalism'. He added in his diary, 'I have become unsuitable for a normal life of civilian orderliness ... Organically, I feel healthy, but somehow used up'.[2]

After his house in Dahlem-Berlin was bombed, he moved with his wife and daughter to a chalet on the Schorfheide, a nature reserve outside Berlin, but was allowed to retain a small office in Berlin until 1944. He called it his Chancellery, and from it he continued to correspond with other ministers. Sometimes ministerial matters were referred to him in error; and this could gum up the works quite badly, as papers landing on his desk simply remained in an ever-increasing geological stratification. Occasionally, staff from the Ministry of Agriculture would descend on him to try to clarify some paper work; an occasion of mutual dread.[3]

This period between his dismissal and the end of the war was the saddest of his life. After 1945, he recounted how rumours were spread that he was mad. 'It is a much greater torture to be persecuted and despised by one's own compatriots than it is to live in the prisons and concentration camps of the victorious enemy ... 'First of all I was considered an idealist, then a romantic, then a rebel, then a defeatist, and last of all a fool.'[4] His diary largely consisted of vituperative rantings between 1942 and 1944, and this section was destroyed *in toto* by his wife in 1970, because she felt it showed Darré in such a bad light. He had joined the great army of the non-men again. As in 1919 and 1926 there was no place for him; this exile was infinitely worse. With all authority lost, his Cassandra warnings went unheeded. The war went as badly as he had predicted, and his reaction to meeting Backe in an air-raid shelter in April 1945 in Berlin was to tell him that he was now justified in his opposition to the war.[5]

Darré gave himself up to the Allies in Thuringia on 14th April, 1945, and was sent to Spa in Belgium, where he wrote a report on the food situation. He seems to have envisaged the Allies appointing him as an interim food minister, and offered many suggestions about the role of the RNS and food production. His first reaction to his imprisonment seems to have been that he could at last be revenged upon his enemies in the Third Reich, and he was eager to co-operate. During 1946 and 1947, he wrote accounts of his activities for his captors and a history of the Agricultural Department from 1930 to 1933. He wrote a short book called *The Stage Management of the Third Reich (Drehbühne)*, which he hoped to publish to help keep his family, who were now being cared for by other members of the family and friends. Hans Kehrl recounts that he shared a cell with him for a while, and described his lively charm of conversation and manner, his infectious enthusiasm for ideas. Another interesting sidelight on Darré in prison comes from the memoirs of a German Jew, Erwin Goldmann, who shared a cell with Darré for a few days, and described him as pro-British and highly regarded by his captors. Goldmann adds that Darré was sympathetic to Zionism, and had helped to train Jews to go to Palestine before the war.[6]

Gradually, he realised the seriousness of his position. His first interrogation, which took place ten days after his capture, revealed a truthful account of his connections with the NSDAP and his experience in office, with the exception of the omission of his membership of the DVFP in 1923, something that was, as mentioned earlier, omitted from his application to join the NSDAP.[7] The picture was disbelieved by his interrogator, who decided that Darré had cobbled this story together in the ten days since his capture. While he was later to be disingenuous about the racial element in Blood and Soil fundamentals, arguing that it was irrelevant in practice, he was in general a truthful witness. He irritated the interrogators by denying knowledge of sinister wartime activities. The judgment of his evaluating officer in 1947 was that he was unreliable and untruthful, a judgment that would have been more convincing if he had been able to spell Darré's name correctly, and if any arguments had been advanced for the judgment.[8]

What does emerge clearly from all his interrogations is the ignorance of the American interrogators of the structure and functioning of the Third Reich. The whole process seems to have been motivated by a vengeful incompetence, in part fuelled by the salacious hate-propaganda of the American Press. Even as late as 1946, the mass suicide of Russians who had fought for Germany and were to be repatriated back to Russia, was headlined by the USA Army paper *Stars and Stripes* as 'Red Traitors Dachau Suicide Described As Inhuman Orgy'.[9] Darré found his hereditary farms described as 'Teutonic Breeding Centres'.[10] He was assumed to have a detailed knowledge of events in the incorporated areas up to 1945, and even when his argument that he had not held governmental authority there after 1939 was accepted, he was pressed for months in 1947 for information about others. His life of guile, manoeuvre, and intrigue, his efforts to cope with hostile Nazi ministers now stood him in good stead, as he was able to counter the threats and blackmail of the interrogators with his old mixture of quibbles about competence, dumb insolence and bored persistence. He also suffered less of a shock from the discovery that he was seen as a war criminal than did many civil servants and soldiers, because he was accustomed to being regarded as an enemy.

The ignorance of the interrogators about Darré's activities was perhaps surprising. They had had a mass of captured documents to show them something of the detailed workings of the Third Reich and the conflicts of authority. Of course, the early information had never been unbiased. Of the very nature of things, little first-hand information could be available to them until 1945, only such intelligence as they could acquire in wartime through refugees and spies. There was plenty of rumour, and the stale coffee-house gossip of capital cities at war. But it is surprising that the Hereditary Farm Law, copies of which abounded, and which was brief and clear enough, was interpreted as a means of stealing land from Polish farmers in Germany, part of the war-time Germanising concept. Darré's lack of real power after 1939 must have become known to the Allies from their captured documents, and indeed, the BBC broadcast referred to earlier had obviously been based on some genuine and personal knowledge.[11] Eventually, Darré became unhelpful to the Allies. Unlike many stauncher Nazis he refused to give evidence against others. His affidavit on the SS for the SS Trial in 1949 (Case 8) was remarkably candid about his racial aims, his decision to use the 'Jewish phenotype' as a negative to make the Germans more racially aware; but refused to enter into speculation about later SS activities, or accuse others.[12] This took some strength of character, after four years of imprisonment and threats of prosecution. Darré's name was 'among those canvassed for inclusion in the Nuremberg Trials, but it duplicated other, even stronger candidates for inclusion',[13] and he was not prosecuted as a major war criminal in the 1946 Nuremberg trial, held under the joint auspices of America, Russia, Britain and France.

But, considering the connotations of Blood and Soil, by now seen as a justification for genocide, Darré was lucky. He defended himself on this issue in his interrogations, arguing that other countries, including America, also had racial theorists, but had not murdered millions; in other words, that there was no connection between Third Reich practice and racial ideals. He referred specifically to an American writer tactfully described by the stenographer as 'G', presumably Madison Grant, to the embarrassment of the American questioners. After Darré had persistently brought up 'G''s name, the question of Blood and Soil and mass murder was dropped.[14]

However, it was to play an important part at the trial. His 1949 trial took place under the aegis of the American Government, and was part of what was known as the Wilhelmstrasse, or Ministries Trials, after the street in Berlin where many of Germany's national ministries had been housed. He was tried jointly with twenty-one others, including civil servant von Weiszäcker, Schwerin von Krosigk, ex-Foreign Minister and Protector of Bohemia, Walter Schellenberg, and Paul Körner of the Ministry of Economics. They were a varied group in conviction, experience and activity, including pre-Nazi conservatives,

Nazi ministers, career civil servants, and Schellenberg, widely regarded within the SS as its cleverest intellectual, and, after Heydrich, one of the most dangerous. Some of the judges had doubts about the concept of a joint trial, and these were raised in a dissenting judgment which will be given in more detail later.[15]

The news that he was about to be prosecuted left Darré in a state of near-suicidal depression. His physical state was poor, he was in and out of the prison hospital, and he was worried about his family, especially during the bad winter of 1947, which saw famine in Germany, and strikes among the starving labour force. His elder daughter came to visit him at this time, and described how much his spirits had been lifted by the presence of his defence lawyer, Dr Hans Merkel. Merkel, who had been called in 1933 from a thriving legal practice in south Germany to join the legislative drafting team at the RNS, was a superficially dry, sober, calm man, with a deeply imaginative streak. He found Darré in despair, and managed to inspire him with hope and determination again, probably, indeed, giving him back the will to live. Merkel saw Darré as a tragic hero of Shakesperean proportions, a King Lear figure. He persuaded him that there was one more chance to justify himself to 'the grandchildren of today's peasants' (as Darré had written in 1939) and to defend his old beliefs and actions.[16] The result was a defence that was remarkably blunt and uncompromising by Nuremberg standards. It defended Blood and Soil and the idea of a peasant Germany in a way that interested the judges, who seemed to be attracted to some of Darré's ideas as well as puzzled by them.

The 'Ministries case' was the longest of all the Nuremberg trials. There were eight counts. Some defendants were charged with all, others with some of the accusations. Count One concerned Crimes against Peace, and Count Two, the Common Plan and Conspiracy, charges which worried one judge because of the difficulties of proving joint conspiracy to wage war, and the juridically novel nature of the notion of a crime against peace. Since all the Allied powers, including America, had in their time waged aggressive war, a doctrine was introduced preventing the German defendants from arguing in their own defence, 'Well, you did it too'. Count Three concerned murder and ill-treatment of POWs. Count Four, atrocities and offences against German nationals on political, racial and religious grounds, was quashed at the beginning of the hearing, and Count Two was thrown out later in the trial. Count Five concerned atrocities and offences committed against civilian populations and where Darré was involved was divided into three parts; first, racial policies and Jewish extermination, second, compulsory expropriation of Jewish land at under market value, and third the compulsory expropriation, enserfing and expulsion of Jewish and Polish farmers in the incorporated areas. Count Six concerned plunder and spoliation in the occupied territories, which in Darré's case meant Poland. Count Seven was the slave labour charge, and Count Eight alleged membership in criminal organisations. The judgment of the International Military Tribunal was introduced as *prima facie* evidence that certain crimes and conspiracies had occurred, and the defence was not permitted to query these findings (e.g., that membership in the *Reichsleitung*, the National Leadership Corps, was a crime).[17]

Compared with the trials carried out in the heated atmosphere of 1945-6, the American-run Ministries case was carried out in a more scrupulous and legal manner. The charges were carefully examined, and as mentioned earlier, two were rejected as having no legal foundation. A dissenting judgment on the judicial status of conspiracy charges and individual responsibility came from Judge Leon Powers, something that would have been unimaginable in the International Military Tribunal (IMT) trial, although it was not read in open Court.[18] The conduct of the trial did present other problems. Men with varied careers and governmental wartime responsibilities were being tried together. Much of the translation was carried out hastily and faultily. Prosecution documents were not made available to the defence until they were read out in court, and often not then. There was little care taken to check the authenticity of the documents. Use of a document in the IMT was assumed to prove its authenticity. Prosecution witnesses did not have to submit to cross-examination from the defence.[19]

Some of the allegations against Darré look bizarre today. For example, much was made of the autarkic nature of the National Food Estate Marketing Law. It was claimed that from 1933 it was a means of preparing for an aggressive war. Such actions as preparing ration books and stockpiling food were alleged to be criminal preparations for a war of aggression, although Britain, certainly, had been contemplating rationing schemes since 1936, on the recommendation of a Parliamentary Committee, and produced ration documents before the Second World War.[20] Charges of plunder and spoliation in Poland, Count Six, omitted to mention that much of the food imported from Poland was re-exported to other areas, such as Austria, Czechoslovakia, Belgium, Norway and France. Food was also exported *to* Poland. An odd impression was given by this omission. There was more food available in Poland during the war than in west Germany's industrial areas, and the peasants and the Ukrainian minority, especially, did well.[21] German administrators in Poland had found a countryside grossly impoverished by their own standards, where the most basic agricultural infrastructural requirements had to be imported from Germany. The idea that Polish factories were solemnly dismantled from these rural areas where peasants were so poor that they had only mud floors and gaping windows was widely regarded in post-war Germany as a fantasy, presumably inspired by the Russian and Allied dismantling of German factories, railways and scientific equipment that took place between 1945 and 1949.[22] Reference in the trial judgment to the German 'pretence of payment' must have rung wryly on German ears, since no such pretence was being made by the Allies after the war.

In the judgment of 11th–13th April, 1949, Darré was acquitted of Count One, crimes against peace, on the grounds that 'he did not attend Hitler's conference where plans of aggression were disclosed', and his one letter to Goering which discusses his plans to resettle ethnic Germans in the east did not provide evidence that he knew there would be an *aggressive* war. Charge Two was thrown out generally. He was found not guilty on part of Count Five, the charge of anti-semitism leading to Jewish extermination, but his attack on 'Jews and democracy' was singled out for comment. The tribunal argued, 'Darré's speeches attack Jews and democracy, but he also attacks the Prussians and Prussianism ... Except in an authoritarian state, it has not yet been suggested that to hold such views is criminal',[23] He described Darré's reference to 'Prussians, Jews and Jewish ideas' in his anti-semitic speeches as 'window-dressing'. The implication seems to be that to ban criticisms of 'Jews and democracy' might lead to a ban on attacks on Prussians and Prussianism, which would never do.[24]

Darré had not played down his anti-semitic statements, but put them in context as political weapons used to defend and further his own position in the 1930s, and stressed his lack of anti-semitism in his two books, published before he joined the Nazi Party. Perhaps the Jeffersonian tone of the ideal of a self-sufficient, racially homogeneous yeoman peasantry came over. Certainly, Darré received a surprisingly sympathetic hearing. 'Some of his ideas were novel and somewhat bizarre, but it is not a crime to evolve and advocate new or even unsound social and economic theories'[25] a judgment that gives some indication of how persuasive Merkel's arguments had been. To allow such autonomy to ideas previously presented as intrinsic to Nazi ideology displays a considerable shift of opinion from the interrogations of 1945.

The compulsory purchase of Jewish-owned farmland at settlement valuations was, however, a clear breach of all accepted legal principles, and Darré was rightly found guilty, although the additional argument that the theoretical difference between the speculative value of the land and its settlement value somehow helped pay for Germany's war effort, remains unproven.[26]

Darré was found guilty of 'callousness but not of criminality' in overseeing the different rations made available to different sections of the population in Germany. The trial concerned itself specifically with Jewish rations, but there were multiple gradations concerning different categories of worker, age group and nationality, not to think of POWs, labour levies, voluntary workers.[27] The most important finding in Count Five was that Darré was guilty of expropriating and reducing to serfdom hundreds of thousands of Polish and Jewish farmers in the course of the ethnic German resettlement, as recounted briefly in chapter Six. The court found that Darré's differences with Himmler were the result of power struggles, not of ideology, and referred in this connection to Darré's letter to Lammers, where he claimed to be better prepared for settlement projects than Himmler's SS men.[28] But the loss of factual authority does not seem to have been sufficiently taken into consideration. The actual letter shows that Darré thought the SS would not understand the special needs of peasant farmers, and he argued that the RNS had overseen internal German peasant settlement since 1939. However, the court interpreted the letter to mean that Darré had been specifically preparing for the Polish resettlement, and it was a major factor in his being found guilty. Despite this confusion, it is obvious from Darré's early works that he did support the aim of German settlement in the Baltic and in former German territories, although intra-German settlement was his main theme. One is bound to ask, *would* Darré have overseen the resettlement of the incorporated areas if he had had the authority to do so, or would he have refused because it was inhumane. Surely the latter course would have been most improbable. Nonetheless, one must refer also to his strictures against von Gottberg's actions in Czechoslovakia in 1938, quoted earlier. Clearly, humanitarian considerations were part of his viewpoint in this matter, if only because he was sensitive to Germany's need for a favourable world opinion. One defence of Darré would be that the Poles had brought it on themselves by dispossessing 700,000 German farmers in the 1920s from the borderlands, but any such point was disbarred from the tribunal, and would in any case make no *legal* sense. Possession, force and conquest are, and must always be, uneasy bedfellows with international principles. My own conclusion is that the finding of guilty made moral sense, even if legal nonsense, as Darré had no authority in the matter.[29]

The question of Darré's degree of anti-semitism in the 1930s is a complex one, and deserves fuller discussion. His circle of Nordic racialists had not been noted for anti-semitism, although Lehmann, his publisher, certainly saw Jews as hostile to the Nordic movement. The author of a monograph on Günther comments that while National Socialism was unthinkable without anti-semitism, anti-semitism was not a valid criterion for Nordic thought.[30] Darré's first two books, which upheld the Nordic idea as a positive racial ideal, did not mention Jews, except in two footnotes in *Das Bauerntum*, where he praised Jewish racial pride. There was a distinct change of content and emphasis on this issue in his articles and speeches between 1929 and 1936, and while sincerity and opportunism are entangled in his work, especially once he became a Minister, it does seem that Darré moved from a position of disinterest in 'the Jewish question' to stressing it between 1931–5, to disinterest again when his political position weakened.

In 1926, Darré had written in praise of Rathenau's racial theories and opposition to democracy, complaining that the Nordicists had no idea what 'the Jew Rathenau' really believed about the Nordic race and the value of breeding, an exercise which clearly meant to tease respectable Jewish non-racialist democrats as much as it meant to provoke the Nordic movement; and two years later he defended himself against attacks by *völkisch* opponents by quoting seven more pages of Rathenau.[31] In 1928–9, he criticised all Semitic peoples, together with Tartars and other nomadic tribes, for their nomadic spirit. He analysed nomads and peasants in terms of a genetically transmitted mechanism for cultural distinctions.[32] In *Neuadel* he plagiarised Günther: 'The real core of Jewry has pure blood, although one cannot talk of the Jews as a separate race in the sense in which racial theory is understood'. Jewish and Polish minorities in the Ruhr were cited in *Neuadel* as examples of non-assimilated Germans. 'It is not always necessary to think of "Jews from the east"; the Polish islands in the industrial part of Westphalia, for example, are also strange to us.'[33] Here Darré was directing his arguments to anti-semitic readers, and trying to broaden their attention. Was Darré engaging in self-censorship in avoiding anti-semitic tirades in these works? This seems unlikely, as Lehmann published many anti-semitic works, and would have welcomed such attacks from Darré.

In his two books, the Jews are not presented as a racial danger or physical threat. After he joined the NSDAP, the cultural-racial dichotomy of nomads versus peasants focused on the Jews as a typical urban and capitalist minority group, Darré subsuming nomadism with capitalism. This dichotomy was used to emphasise the supposedly non-mercantile nature of the rural German. But how could a cultural feature, such as the inherent dislike of urban life by true Germans, be transmitted genetically? He suggested that while there was as yet no scientific explanation for this peculiarity, it probably lay in the different ways nervous systems reacted to the environment. He argued that Jews *liked* urban life, and that their breeding rate remained as high in towns as in the country, which proved they were natural city dwellers. Following up his theory, Darré examined different behaviour patterns between domestic rats and wild rats. He claimed that domestic rats were affected unfavourably by the environment, and ceased to breed, while wild rats were unaffected, and continued to breed, and for that reason, wild rats were used for experiments by scientists. There had to be a genetic mechanism at work which controlled this group behaviour. While there is clearly a pejorative element lurking behind Darré's identification of Jews with urbanisation, it is not rendered programmatic.[34]

Similarly, Darré in 1927, in the colourfully titled 'Pig as Criterion regarding the Nordic and Semitic Peoples' [*DE*, 3, 1927] used a real cultural distinction—dietary differences between different peoples—and rational argument to reach his suggested conclusion. He rejected climate as a cause, and suggested that religious prohibitions were effects not causes. The reason Semites hated pork and Nordic peasants loved it, was that the pig was a symbol of the settled way of life. They were hard to rear by nomads, and thus disliked, and this dislike gradually became a dietary prohibition to symbolise an anti-agricultural attitude. Interestingly enough, a similar argument, obviously arrived at quite spontaneously, is used in a recent work on the history of food, where the author gives the same reason for dietary prohibition of pork among Semites: she argues that the dislike of pigs was borrowed from nomadic *Nordic* tribes, who themselves dislikes pigs because they abhorred agriculture.[35] Darré included Tartars, Arabs and other eastern peoples in his nomad category, not singling out Jews for special attack.

The lack of anti-semitism in Darré's two major works attracted unfavourable comment from Hitler in 1930, comment which was passed on to Darré when he was angling for a meeting with him, as described in chapter Four.

> A.H. has been falsely informed over my *Bauerntum* insofar as he believed that I didn't sufficiently interpret the Jewish problem in terms of their parasitical essence. Briefly and clearly, you can see from these few indications how things stand.

At first sight, this seems straightforward enough. Hitler thought he was soft on the Jews, but Darré, in describing this criticism, takes care to describe them as parasites, so clearly he is *not* soft on the Jews. But after all, Darré had certainly not sufficiently 'interpreted the Jewish problem' in his books. Read in the context of the conspiratorial correspondence with Kenstler, it may be that Darré was preparing for the public shift in attitudes which would enable him to join the Nazi Party in specifically repudiating the allegation that he failed to make the Jews the essential core of the world's problems.

From 1931, he associated 'Jewry' with rootless, urban capitalism, which he defined as a system of exploitation and plunder, deriving from the landless nature of nomadic tribes, and their habit of conquering and plundering others. In an attack on Damaschke, he alleged that the Land Reform movement was Jewish-inspired, and represented the re-emergence of the old antipathy between peasant and nomad. Land Reform was inspired by Ricardo, a Jew. Ricardo's theory of rent attacked land ownership, and was taken up by anti-peasant Jews and pacifists. Nineteenth-century land reform was collectivist, and called for land nationalisation, 'the typical nomad dream, drones who would live off the produce of the man of the soil-parasitical'.[36] Of course, the attack on land nationalisation was not made from a property rights standpoint, and his own proposals assumed strong collectivist measures. But he differentiated his own socialism from 'Jewish socialism' by claiming that his was racialist, benevolent and patriotic, while Damaschke was inspired by a malevolent hostility to the Nordic peasant, and a desire to infiltrate right-thinking groups. This attack was clearly opportunistic, in that Damaschke, while inspired by Henry George, was a co-founder with Naumann of the *Nationalsozialer Verein*. In many ways, he was a typical social reformer, teetotaller, *völkisch* figure of the period, and is described by one historian as anti-semitic.[37]

In late 1934, Darré wrote of the 'counter-concept of Jewish nomadism'. He ascribed the peasant wars of 1525 to 'the eternal representative of all international finance capital, the *Jews*'. International Jewry attacked old peasant freedoms. The Jew was both nomad and trader. Profit was inherently opposed to the peasant ideal.[38] The peasantry was presented as an anti-mercantilist bastion during this period up to 1935: Darré's circulars to the National Peasant Council, a body of high party dignitaries, usually contained anti-semitic arguments. They served to stress his absolute soundness on this key issue, while enabling him to present the peasantry as the *Stand* most worthy of Hitler's support. By 1936, Darré was enunciating classical conspiracy theories.[39]

Of course, like other Nazi ministers, Darré was subject to pressure from below, both from the peasantry and from the Nazi activists, who, in 1934-5, complained that Jewish traders could join the RNS, and operate in German agriculture.[40] None the less, pressures against Jews as a racial group, as opposed to the idea of a Jew as representing unpopular market forces, were weak in rural areas.[41] Although the Jewish cattle dealer was a stock target for anti-semitic peasant attacks from 1880 on, in practice peasants sometimes protected Jews from Nazi attacks. As late as 1940, it was only under strong pressure from Kaltenbrunner (later head of the RSHA after Heydrich's assassination) that Jewish-owned land was finally bought-out compulsorily. Correspondence between Backe, the then minister, and local officials, shows the programme's unpopularity, while there was massive foot-dragging by officials in formerly Austrian areas.[42]

Darré's circular to his staff tried to clarify the confused legal situation on the employment of Jews within agriculture after the Nuremberg Laws of 1935.[43] But he called specifically for anti-semitic boycotts to cease, while calling for placards to be placed in town centres proclaiming, 'Race is the key to world history'.[44] This was more provocative than it seemed. He was arguing that hostility to Jews was not only inhumane, but encouraged a materialistic, envious attitude among the perpetrators, based on an inferiority complex, not on a racially positive ideal. Not only was this playing with fire in the atmosphere of the time, but the words suggested were even more dangerous, as they were the well-known quotation from Disraeli's *Coningsby*.

The picture that emerges from the contemporary documents is certainly not that Darré was in any way pro-Jewish, but that he did not want his own pro-Nordic programme to be absorbed by an emphasis on negative anti-semitism. He encouraged 'positive racial education' by publishing pictures of healthy young gymnasts. In 1934, embarrassment arose when a half-Jewish cover girl modelled the cover of *Neues Volk*. The girl, a Viennese actress, left for America before the mistake could be detected.[45]

Darré also supported agricultural training camps for Jews to go to Palestine, and told local National Food Estate officials not to bother about Jewish ancestry if it meant that good peasant stock would be excluded from becoming honorary German farmers under the *Erbhofgesetz*.[46] So was he a closet anti-semite who came out when it was safe to do so, or did he adopt a public stance against the Jews to shield a weak position? He was certainly influenced by friends like Rosenberg. When Hore-Belisha was appointed Minister of Defence in Britain, Darré wrote in his diary, 'Now England is a Jewish instrument of war'. He forwarded an article published in a French journal to Hitler which discussed Hore-Belisha's allegedly Moroccan Jewish parentage. The diary entry would hardly been part of a public stance.[47]

On the other hand, Darré was not perceived by others as having persecutory feelings towards Jews. The Hereditary Farm Law, while it excluded Jews from the role of honorary German peasant, did not itself interfere with specifically Jewish land ownership. Perhaps his private outbursts represented a generalised and endemic anti-semitism of the time, rather than an accord with the inner circle of Nazi anti-semites.

Darré was also found guilty of plunder and spoliation in Poland, although here, too, his lack of real authority at the time should have been taken into account, since this lack was not seriously disputed by the tribunal. But his successor, Backe, was no longer there to try, and public opinion demanded a victim. Bearing in mind the sufferings of the victorious powers, this was not surprising. The Russian government, in particular, needed a scapegoat to blame for the miserable condition of Poland, now undergoing an undeclared and secret civil war, and the hunger and spoliation of the eastern borderlands, largely caused by Russian looting and atrocities. In this charge, as in many Nuremberg allegations, one seems to see a mirror image of many of the post-war problems.[48]

The last serious charge was that of using slave labour for agriculture. Darré was found not guilty, on the interesting grounds that 'it does not appear that at that time (1940) there was any demand for forcible recruitment by the agricultural authorities, nor that any action was taken by the General Council of the Four Year Plan for such forcible recruitment', and that there was no satisfactory proof that Darré 'ever suggested forcible recruitment'.[49] This was an understatement, indeed, since he had vehemently argued against such recruitment.

Darré was cleared of Count Eight, I, of being a member of a criminal organisation, the SS. In view of the fact that he had been intimately involved with the SS between 1931 and 1938, as well as being a member, this may seem surprising. However, the fact that he had attempted to resign in 1938, and had been kept on the rolls against his will, and that he had resigned his office in the Race and Settlement Chief Office, was taken into consideration. The judgment seems to imply that the SS was not a criminal organisation before 1938. He was found guilty of being a member of the National Leadership Corps.[50]

The sentence was seven years, and consideration was given to the four years already spent in jail. Darré was fortunate not to have been tried in 1946 along with Streicher, Ley, Goering and other leaders. His sentence was lenient, especially since he did not benefit from the growing realisation of the true nature of the Russians by the other Allies, as did the Wehrmacht officers and industrialists.

Certainly, Darré was lucky to be found not guilty of charges of anti-semitism, but unlucky to be found guilty of dispossessing Polish farmers. Both were really miscarriages of justice. Assuming anti-semitic statement to be a crime, he was guilty.

Between 1947 and his death in 1953, Darré received hundreds of letters addressed to him as 'Herr Minister', or 'Dear and Honoured Minister'. Indeed, one acquaintance commented that imprisonment was in a sense a boon to him, as it restored his lost status. As far as the Allies, too, were concerned, he was once again the Minister.

This support from outside also helped to lift his morale, and was the culmination of a long process of returning mental health. Bad as the post-war period was, it was the awakening from a bad dream for him. Now, gradually, a world of normality returned, precisely because the catastrophe he feared had occurred. A beaten Germany lay starving and dismembered, with millions of refugees homeless. His own dependents, like many others, were penniless, but at least matters could not become worse.

Darré was freed on appeal, largely on the grounds of ill-health, and because he had spent a total of five and a half years in prison already, after a hearing in Munich, August 1950. He was one of several convicted Nazis released that year, including Otto 'Press' Dietrich, and Flick, the industrialist. In order to qualify for release, the prisoners had to have been given fixed term sentences, not life, and to have served a certain proportion of them, roughly one third, with good behaviour. The releases caused fierce criticism in the *New York Times*. 'Democratically minded men and women' might, it was feared, turn against the Americans because of their leniency. Fortunately, America managed to survive this threat. English papers noted worriedly that delegations of East German trade unions (possibly the democratically minded men and women mentioned by the *New York Times*) also opposed the releases.[51]

He was released in October, and went to stay with his former in-laws near Stuttgart, where he walked in the countryside, and reflected on his experiences. His letters express a more fervent love of the German countryside than he had ever before articulated. His first action was to write to Merkel, thanking him for all his efforts on his behalf.[52]

In Stuttgart, he met an organic farmer, a meeting arranged by the former secretary to the Anthroposophist Society, and explained his plans to start a German version of the English 'Soil Association', the group, still in vigorous existence, which Lady Eve Balfour had started in 1941. A publisher had offered him a study to work in Goslar, and he accepted the offer gratefully, feeling that this renewed link with Goslar, where his mother was buried, and where he met his first wife, must be fated. He planned to settle in Bad Harzburg, a spa town near Goslar, and live on his earnings from journalism. 'In any case, I have the feeling that, although things look bad in material terms, there is some hope', he wrote, comparing 1950 to the dark days of 1930. The irony of such a comparison, in view of what happened during the later 1930s, seemed lost on him.[53]

In fact, although bombarded with Steinerite material while in prison, Darré never became fully converted to any of these ideas. His peasant ideology lacked that degree of utopian fervour: it was based, rather, on Darré's vision of the most sensible, practical and fulfilled way to live. He followed T. H. Huxley's view on the essentially artificial nature of civilisation; man had become a domesticated animal, and needed careful cultivation to maintain a viable society. This entailed not only eugenic and intra-racial ideas (German peasants being deemed to be German already) but environmentalist improvements in infant nutrition, gymnastics and fresh air for the young, and other pro-natalist, neo-Lamarckian ideas. Shortly before his death, Darré wrote to a young follower of Steiner who proposed to farm a smallholding on organic lines. He began by encouraging the general aim, which he saw as a pioneering goal comparable to nineteenth century settlement abroad. He went on to disassociate himself, kindly, from the more extreme views on 'biological-dynamic' farming, then argued that over the centuries man had suffered the loss of two kinds of relationship: 'the relationship to the organic strength of the soil and earth, and secondly the relationship to God, who lives and works in everything in this world'. Prior to the appearance of 'materialistic philosophy', the peasant had been the link between man and God. This holistic trinity, peasant-nature-God, was the only way for man to fulfil himself. Darré added that he had been a fool to think that the Nazis were the ones who could repair this link.[54]

Rolf Gardiner resumed contact, writing first to an ex-member of the RNS. Several letters were exchanged between him and Darré, and Jorian Jenks, ex-agricultural adviser to the BUF and proponent of smallholdings, probably on Gardiner's request, sent Darré his writings. But Darré was cool. Gardiner's generous and selfless gesture in offering friendship to a man who was, after all, a convicted Nazi war criminal, was not really appreciated. His attack on the RNS and, to some extent, Darré, in his broadcast in 1940, after accepting German hospitality during the 1930s, had never been forgiven. The radical Nazis had no idea of the atmosphere in wartime Britain, and the strength of public opinion.[55]

Admiral Horthy who had encountered Darré when the two were imprisoned at Spa, Belgium, in May 1945, tried to meet him when he passed through Berlin, but he was ill in a clinic in Berlin, and unable to keep the appointment. Thirty years ago, Horthy had distributed land to his veteran soldiers, and Darré had alluded to this in *Neuadel*. Horthy may have heard of it through the Hungarian veterans who had contacted Darré in 1930 after the publication of his book.[56]

Another move to form a German 'Soil Association' was made in December 1952, when Darré met the Town Clerk (Oberstadtdirektor) in Goslar at the Hotel Niedersachsischer Hof, and made notes about a society to be called 'Mensch und Heimat'. Its function would be to further 'organic ideas, a healthy soil and care for the homeland' (*Heimatpflege*). The society would have to be officially registered and incorporated, and much of the discussion turned on how this was to be done. Darré could find financial support from his brother-in-law, and legal backing from the ex-RNS lawyers, possibly Merkel, but was perturbed about his own involvement. 'How to avoid the accusation of being a "naziistische" organisation, while retaining some influence,' he jotted.[57] During the years after his release, he wrote steadily, articles with titles like 'The Living Soil', 'Peasant and Technology', and

'Mother Earth'. The articles on organic farming were usually inspired by English works, such as those by Sir Albert Howard, Sir George Stapledon, and Lady Eve Balfour, although he also referred to the USA's 'Friends of the Soil', and American efforts to combat erosion. In 1953, he enthusiastically reviewed Lady Eve Balfour's *The Living Soil*. As in the 1930s, he wrote on the American dust bowl, this time under the pseudonym of Carl Carlsson, and called for Germany to adopt soil protection measures of a similar kind to America's 1947 anti-erosion law, in the local press, and in a journal published by an old Witzenhausen friend, in South America.

His articles attacked large corporations for ignoring the possibilities of new technology which could help the farmer. He opposed exploitative attitudes to the land, factory farming, and an 'agricultural division of labour', all caused, in his view, by contemporary technology. He especially criticised industry and government for their obsession with the virtues of large-scale farming, and demanded encouragement for machines which would help the peasant farmer. As in the 1930s, though, he hailed the development of the small, petrol-driven tractor and machinery useful for market-gardening. He energetically propounded the virtues of methane gas plants for small farms; such plants could produce energy from composted dung and other waste materials.

By the summer of 1953, Darré was too ill to continue. An appeal was launched to pay for his medical care, and among those contributing were political enemies who none the less respected Darré's commitment to the agrarian sector, men like the head of the Bavarian Farmer's Union, who insisted on paying something out of his own pocket. He went into a clinic in the Leopoldstrasse, in Munich, helped by his sister-in-law and long-term close friend, Karen von Billabeck. He was soon unable to write, and scribbled an apologetic last note to his first wife, Alma. There he died on the 6th September 1953 from liver failure exacerbated by jaundice. His death was marked by some guarded obituaries in Germany, but aroused little interest.[58]

During the war, Darré liked to walk on the Steinberg, a mountain overlooking Goslar, where there was a peasant leaders' school. Attendance here for a year was compulsory, and they owned a field, where Darré planned after the war to build a centre for farmer's wives and their children, with a creche, a library and sports facilities. The field in Goslar now boasts exactly such a building, erected in the early 1960s by the local farmer's union, and looked after by a former peasant leader, but it holds no reference to Darré, and any trace of Darré and the National Peasant City have long since disappeared from Goslar. The hotel, Der Achtermann, where the mediaeval basement was used for RNS rallies, was closed for decades, and only recently opened, now under new ownership, and with largely Turkish staff. The town itself found its population doubled by refugees from Thuringia, in the Russian zone, and refugees from Poland. It is hardly surprising that Goslar's monument to Blood and Soil, with its wheat sheaves, its Odal rune and the swastika, was dismantled; perhaps unsurprising that after Darré's death, the Christian Democrat mayor of Goslar, exiled by the Nazis, tried to prevent his burial in the town cemetery, where he had long reserved a place next to his mother. It was the Social Democratic Town administrator, who shared Darré's interest in the organic farming movement, who insisted that he should have the place. But both men attended the 'honorary citizen's' funeral, as, according to the local paper, did hundreds of Goslar citizens. Dr Winter, who had written to King's College School, Wimbledon, on Darré's behalf in 1933, laid a wreath from the German Colonial School at Witzenhausen.[59]

The gravestones have long since been removed from the mounds of earth where the bodies of other members of the National Peasant Council, based at Goslar, were buried, but Darré's grave retains its stone, a large plain slab under whispering pine trees, in the corner of the secular cemetery where Guderian and other military leaders are buried. There is no name on the stone, but an allegedly family shield, bearing a strange resemblance to the Nordic quasi-religious symbols of the 1920s—enclosing the pagan symbol of eternity, a snake with its tail in its mouth, carved around a runic S with a bar through it—and the withered memory of twelve red roses cast on the grave by his sister in 1953. From the stone, it is possible to look up to the Steinberg, where the Peasant House is just visible.[60]

Before 1945, each small town had its 'Darré-Haus'; now one only is rumoured still to have the name on the door, in East Germany. Goslar typified Germany's attempts to wipe her memory clear after the war. It took eight hundred years before the old imperial palace was rebuilt in the nineteenth century: one hopes it will not be so long before the brief episode of Goslar, the National Peasant City, can be exhumed again, examined and remembered.

In both of his main points, the need for man to be linked to the soil, and for the maintenance of the vitalist cycle of soil-growth-decay, Darré was directly in the tradition of much of today's 'Green' thinking.

In his belief in a peasant international, stretching from England to the Baltic, he expressed that element of anti-expansionism and anti-imperialism which characterises writers from Cobbett to Goldsmith who have propounded the peasant ideal.

It may be argued that it is impossible to take seriously any ideas put forward by a minister of the Third Reich. Darré was a racialist; does this not put his ideas beyond the pale? Can his writings after the Second World War be taken seriously when it is most unlikely that he would have been allowed to publish works on the virtues of Nordic Man? Can his interest in organic farming be anything other than window dressing for support for a vicious, brutal regime?

Of course, it is unlikely that Darré would have been able to publish book reviews, even under a pseudonym, propounding the survival of the Nordic race in 1951. However, as early as 1930, he seems to have put his Nordicism second to his peasant ideals, and was certainly criticised by members of the Nordic Ring for doing so. But in any case, intra-racialism is an intrinsic part of the peasant ideology, although it is unfashionable to stress this in the west today. The Third World and ethnic minorities in the west are more openly racialistic. The idea of a network of kinship is intrinsic to a definition of a peasant, and must by definition exclude those alien to that network. The ideal peasant farm is oriented towards the long term, the family and the future of the tribe. 'Peasantness' cannot absorb alien cultures, religions and races, without the risk of self-destruction. Further, in Darré's case it is hard to criticise his sincerity, when he constantly put himself in danger of imprisonment or worse, and saw many of his friends and allies arrested for holding his own views, but none the less persisted with those views.

One returns to the paradox referred to in chapter Eight, that if National Socialism, as an ideology, had room for Darré's views, then it may be that currently accepted categories of early National Socialism will have to be revised to allow for the sincerity and, indeed, passion, with which a top Nazi leader held views on the environment, the ecology and the peasant. If the view is accepted that Darré's peasant ideology presented merely morphological and tangential resemblances to National Socialism, but is an immediate ancestor, in spirit as well as in detail, to the ecological movement of today, then the implications of current ecological ideas may have to be examined.

The continuity of his ideas, from the 1930s to the 1950s, is striking. By the early 1950s, the idea of self-sufficiency in energy and raw materials had become widely accepted. The Common Agricultural

Policy was to continue price support for small farmers, and to some extent, protection for smallholders. It was widely perceived among German farmers as a continuation of the old controlled marketing system, and welcomed for the same reasons. German émigrés in England, like John Papworth and F. E. Schumacher, popularised 'Small is Beautiful' ideas in a post-war Britain still dedicated to fashionable large-scale planning. Eventually, the anti-capitalist implications of the conservationist movement were recognised, and it became politicised. The old organic farming movement, with its innovatory search for new grasses, compost-making methods and self-sufficiency allotments, owing so much to German vitalist thought of the early decades of this century, became swamped in a vaguely 'alternative' movement, egalitarian, anti-nuclear (illogically, since nuclear power was cleaner, safer and less wasteful in terms of resources), and messianic. From re-cycling bottles and saving whales, conservationists became associated with feminism and other forms of exclusivist hysteria. When one is discussing the element of continuity in Darré's ideas, this break in the 1950s should be borne in mind.

One should, in short, perhaps distinguish between today's party political Greens—Levellers rather than conservationists—whose prescriptions are often as unrelated to environmental issues as they are draconian, and the more dedicated organic farmers, whether in Germany or England, who pursued an unpopular and often financially unrewarding path over decades, at a time when their ideas were not at all fashionable. The latter group seem to be dedicated to many of the same values as Darré and his contemporary ecologist and Anthroposophist sympathisers: the Greens bear more resemblance to those pre- and proto-Nazi groups that sprang up during the 1920s in an outbreak of quasi-religious prophecy and radicalism. Here there is room for research. But certainly it is time the contribution Darré and his followers made to twentieth-century ecological thought was recognised; it is at least arguable that without him the ecological movement would have perished in his time and place.

Conclusion

Despite the covert nature of his role in joining the NSDAP, Darré was virtually the sole creator and representative of agrarian theory within it in the early 1930s, recognised as the master and arbiter of all agrarian matters by his peers. His themes were: intensive peasant farming, localised autarky as a step towards national autarky, defensive and eugenically-oriented racialism and a defensive racial nationalism. His profound anti-urbanism was a source of serious and eventually fatal conflict with the NSDAP, with its strong urban, lower-middle-class basis; and only his ability to attract peasant support, and his revolutionary, yet non-Marxist stance, made him palatable to the non-agrarian Nazis. Many serious agrarian experts and reformers approved of Darré's social and racial rationale for supporting the peasantry. Primarily, however, they saw the peasant movement as the most sensible and efficient way to develop German agriculture. After Darré's fall, his ideas were retained by the middle and lower-level agricultural administrators, but with more emphasis on their immediately practical aspects. The populist element in his peasant corporation was also a powerful driving force for those SS staff concerned with ethnic German resettlement during 1940–1.

Between 1933 and 1938, many of the economic aims of National Socialist agrarian policy were successful. Productivity increased in nearly all sectors; national self-sufficiency increased, as did intensive peasant farming: these were all major policy aims. Economic failures came later, and were a result of the low priority given to agriculture during the period of war preparations. Later, war itself, and the misconceptions about possible food production in Eastern Europe, led to further neglect of German agriculture. The success of agricultural reform in the 1930s came about despite a multiplicity of overlapping and conflicting organisations, and the experimental problems of the marketing organisation. It was largely due to the relentless efforts of the middle and lower-level echelons of the RNS, together with the unremitting labour and ingenuity of the German peasant.

Given the limited time-span in which the National Food Estate was effectively operative, and given the conflict between the drive for higher production and the need to protect farmers from the gluts and price collapses of earlier years, the achievement of the agricultural advisers in producing a self-sufficiency level of 81% by 1938 was considerable. This is especially so since after 1936, artificially high producer prices – that most potent stimulus—became politically unacceptable. Wartime food production, too, was maintained as efficiently as was possible in the circumstances, and without the disasters of the First World War. That this occurred despite the interference of Goering, Himmler and the Wehrmacht High Command, the shortage of foreign reserves, machinery, fertiliser and government expenditure, is again a tribute to the competence of the RNS and RMEL administrators—especially since many of the countries occupied by Germany during the war were food deficit countries. As the organisation of food production to cope with blockade and war had been another policy aim, this too must be counted as a considerable success.

During the war, unrealistic assessment in Berlin of the time, money, energy, machinery and vast amounts of fertiliser needed to render derelict and virgin Polish land fertile, was a constant hindrance; as was the obsessional belief among many top Nazis that free land existed 'over the green, green heath'. It became painfully and rapidly clear to the agricultural advisers and administrators sent to the Incorporated Areas and places east that such land needed expenditure on agricultural infrastructure—roads, electrification, transport and housing. This was without the confusion and expenditure necessitated by a complete population transfer (as envisaged, if not fully implemented). Whatever Hitler's true motives for war, the Second World War was hopelessly misconceived in terms of food surpluses and living space, and Darré's pessimism about its results justified. Here, as elsewhere, his capacity to take a long-term view was vindicated by events, despite his reputation for unwordly impracticality. Wartime Europe did remain a food deficit area, and the RNS could never have changed that fact in the short term, because of the rogue factor of Russia. Russian agriculture lay in ruins, its grain-growing area devastated by collectivisation, well before Operation Barbarossa; it seems, indeed, to have remained in that condition to this day.

However, the question remains: given a less expansionist régime, would Darré's social policies for the peasantry still have been doomed to failure? Could the hardships inherent then in German peasant life have been ameliorated sufficiently to stop and reverse the flight from the land? The improvement of peasant living standards was a pre-requisite to any attempt to select a racial élite from among the peasantry, as without it, peasants would drift away from the land. Insofar as his ideas entailed the economic and social protection of the peasantry, there seems no reason to suppose that these features of his policies were anything other than viable. Indeed, the increase in small farm productivity that took place in the 1930s provided the perfect justification for such protection. But Darré's inability to distinguish—at least in his rhetoric—long-term plans from short-term necessities, hampered progress in areas where progress could have been achieved. It enabled critics of the rural ideal, in Germany as elsewhere, to dwell on the peasantry's supposedly anti-modern, anti-industrial, past-oriented and defensive tendencies, and to ignore the mounting evidence that peasant farming was a practical approach to the problems of securing the national food-supply.

Would a less fundamentalist approach have won concessions from the régime, or was there a real incompatibility between the demands of agriculture and the industrial expansion of the 1930s? Certainly, Darré saw peasant society and urban industrialism as mutually exclusive categories, and this was inherent in his vision, but was this division true in real terms? After all, the policy of encouraging increased peasant production was successful. Increased mechanisation, fertiliser use and altered product mix was a realistic strategy for the survival of the small farmer in an industrialised society. Thus far, the small farmer and the urban world were complementary rather than opposed, where economic needs were concerned. On the other hand, while the rural ideal prevalent in the inter-war years was not especially past-oriented, stressing rather the need to break free of the shackles of a burdensome and oppressive past, it was certainly defensive. It envisaged an economically static society, and this, together with Darré's doom-laden scenario of urban decay, the choice of death or a painful survival, made it impossible for most Nazi politicians to support Darré's version of the peasant society. Even his followers, however sympathetic, had become reluctant to accept this vision by the 1930s, for the apocalyptic nature of his—and Weimar—prophecies was inappropriate to the spirit of industrial expansion and technocratic success that ruled Nazi Germany. In the face of the furious rate of change of German society and industry in the 1930s, Darré's society of stasis—an essentially defensive concept, opposed to economic growth—became impossible to maintain as an ideal, except in the very longest of terms.

However, there was an alternative inherent in Darré's vision. This was the strong, productive, creative peasant, the 'smallholding technician of the future'. Such a figure could co-exist with industry, protected by the strongest possible weapon, the economic power of his productive capacity. This was the element of Darré's ideology that survived, and indeed, the decision to use price support and subsidy to favour the peasant rather than the large landowner was a major change from previous German policy, and one whose political implications were considerable. Further analysis of these implications is difficult, and perhaps redundant, since the loss of the Junker heartlands after 1945 would in any case have rendered this political change inevitable, and any rearguard action impossible.

The political implications of Darré's policies are more complex. Peasant populism was a strong factor behind the agrarian legislation of 1933, and a major impulse in Darré's support from the peasantry. However, this peasant populism was inherently hostile to large-scale institutions, especially those associated with the modern nation-state. Agrarianism was symptomised by a republican radicalism opposed to hierarchy and supra-tribal organisation. It had no room for a political system that was not flexible, organic and spontaneous. While agrarian political groups were capable of maintaining their bargaining power within a modern industrialised state (and have done so to this day), the spirit of their self-image, the heart of the rural ideal, tended towards anarchy, a phenomenon noted during the peasant revolt of 1928–9. All normal political forms were seen as enemies, a suitable target for peasant cunning, and that untrammelled violence typical of their unvoiced, centuries-old sense of alienation and oppression. Not only did the attempt to create a corporatist-cum-syndicalist structure, the *Reichsnährstand*, contradict the spirit of the centralised National Socialist state, but as a means of giving the farmer political power, it involved a self-contradiction: peasant populism was inherently anti-state, but needed state legislation and protection to co-exist with the urban world it nourished.

Further, the racialism inherent both in the web of peasant values and Darré's specific pan-Nordicism was hard to reconcile with nation-state boundaries; peasant tribalism was smaller-scale, while Nordic racialism envisaged a Northern European 'green' union, spreading from Holland to Finland, in a loosely knit federation. But this had little to do with the aims of the Nazi Party, which, like most other German political parties between 1918 and 1933, was committed to restoring German power and territory to (as a minimum) its pre-First World War position. Darré's lack of interest in this aspect of foreign affairs would alone have unfitted him to be a Minister in the Third Reich.

But it perhaps unfitted him for a post in any government committed to survival and expansion. The internal dynamism of the Third Reich seemed to lead inexorably to empire, for reasons which are still the subject of vigorous dispute, and which are outside the range of this work. This imperial drive in turn entailed a fully fascist, 'from above' power structure, with an essentially non-racialist, but élitist social base, which crossed national and racial boundaries in search of allies and ability. Empire meant a strong central government, playing off some ethnic and national groups against others within it. No empire had ever survived without a cross-current of foreign influences, of exogenous cultural inputs. Darré's hostility to conservatism and imperialism was caused by his assessment of this inevitability, and the extent to which war and expansion would be to the disadvantage of the peasant; while, almost alone among Nazi leaders, he realised, and protested against, the racial implications of importing foreign agricultural labourers, while German peasants were recruited for death on the Russian front. In practice, Darré was obliged to subordinate his planned racial selection, and the breeding of a peasant élite, to the more pressing need to preserve the existing peasantry. This subordination of racial theory to political practice was characteristic of the Third Reich, not only in the areas of peasant settlement and ethnic German resettlement, where racial rhetoric gave way to more immediate imperatives, and not only under the exigencies of war; it was observable throughout its governmental practice. Even within the SS, supposedly the heart of the racial selection process, the practice of mass co-option of outside advisers and bodies meant that racial selection was secondary to the creation of an 'open, yet authoritarian élite' (Struve) except at its lower levels, where racial selection was used to weed out the politically uncommitted after 1933. Certainly, the SS élite eventually became pan-European, losing even its national as well as its racial character.

If anything, agriculture displayed less inherent conflict between achievement-orientation and racial selection than elsewhere in the Third Reich; not because agrarianism was necessarily more racialist, but because racial selection of the peasantry was not an essential prerequisite of Nazi agricultural policy. Peasants, in a sense, were pre-selected. The vast majority of German peasants were Germanic by definition and by self-assessment. It was *intra*-racial selection, Darré's plan to breed a new nobility, that rapidly took a back seat after 1933, and as the peasantry continued to decline in numbers, it disappeared as a policy goal. The fact that Darré and his old friends among the 1920s *littérateurs* such as Schultze-Naumburg and the Prince zur Lippe, took up eugenic plans in conditions of semi-secrecy in 1939 (as part of Darré's Society of the Friends of the German Peasantry) demonstrates the peripheral nature of this plan where government policy was concerned. The mass importation of foreign agricultural labourers was the final acknowledgment by the Third Reich that the German peasant's role as racial source material was to be abandoned.

But if Darré's racialism does not of itself create a coincidence with National Socialism, how is it to be classified? The implications of treating man as a biological entity had aroused controversy long before the rise of National Socialism, and the disagreements that emerged had little to do with the conventional right-left political spectrum. The belief that mankind was a part of the natural order, and subject to all the physical laws that emerge from a study of animals, opposed the idea, linked sometimes with Utopian, sometimes with Idealist or religious tendencies, that man was unique, and could not be subject to the same kind of classificatory process as the crested newt; a view expressed in the 1920s and 1930s not only by liberal and democratic opponents of the Nazis, but by German neo-conservatives, such as Spengler, Jung, Jünger and Moeller. The view that supported racial typology was held by (among others), left-wing and progressive writers, such as Carl Vogt (who taught Houston Stewart Chamberlain at the University of Geneva); it accompanied a belief in scientific rationalism. Racial categorisation was not unique to National Socialism, nor was it a right-wing or conservative phenomenon. The racialist element in agrarian and pro-peasant thought does not necessarily label it, therefore, as National Socialist.

What is at first sight closer to National Socialist and allied movements is the 'naturist' position, one end of a political axis where the determinant is the posture taken towards the limits of human adaptability. Darré was a naturist. He saw man's need to remain part of the natural world (natural in style, texture and spirit) as an immutable datum, dominating but not destroying the potential of the human will. This immutability of man's nature was not the immutability of the unchanging natural order inherent in the conservative position, but the product of a more radical stance. This radicalism, the mixture of voluntarism and determinist pessimism, resembles the proto-fascist movement immediately prior to and post the First World War; such movements, however, were not involved with agrarian arguments, which remained Darré's primary concern, so the resemblance does not really help to classify the agrarian movement under Darré in the 1920s and 1930s. A forceful argument was made by Nolte that the crucial drive behind National Socialism was the opposition to 'transcendence'. He defined this as the belief, common to liberals, Marxists, and most post-Enlightenment thinkers, that the boundaries and possibilities of nature could be transcended, that man's unique abilities gave him the opportunity of unlimited progress and adaptability. Fascism was an attack on transcendence, defined by Nietzsche as the great crime against Western culture, one inherent in its very origins. Darré fits this antitranscendence definition. But Nolte is surely mistaken in confining this political axis to National Socialists on the one hand and liberals plus liberal-derivatives on the other. Naturism versus transcendence is a division that can be traced far back in European thought; and these categories do not fit a fascist-anti-fascist definition, nor, again, a right-left one.

Where the analysis can be applied to National Socialism—bearing in mind that it in no way applies to Nazi practice—is to the attack on materialism and exploitative technology made by diverse Nazi writers, including Hitler. They presented the spirit of exploitation and maximum utilisation of resources as a crime against nature, usually capitalist and Jewish inspired. However, the degree of involvement in, and reliance upon these arguments is a limited and trivial factor in National Socialist thought, with the exception of Darré, who carried the mechanophobe argument to its logical conclusion. Darré's obsession with the vision of man as a natural animal, his belief that the peasant was the link between 'a Holy Trinity of Peasant, soil and God', had no room for extraneous inputs, such as loyalty to Hitler, or to the Nazi's watering down of the racial ethic to assist a perceived national-cum-imperial requirement. He combined practical arguments about peasant production with impractical demands for a revolutionary social purification, a near-nihilistic rejection of existing institutions, demands which could not possibly have co-existed with the NSDAP as it emerged after 1933.

We are left with the man who created an agrarian policy for the National Socialists, and brought into the NSDAP a large and significant section of popular peasant opinion and intellectual support. His fall left a joint legacy of disaffected farmers and devoted agrarian reformers, who continued to try to implement what they conceived to be his aims. After the Second World War, more of Darré's reformist ideas were implemented, in the shape of the peasant protection mechanism enshrined in the Common Market Agricultural Policy. The Messianic underpinnings of his holistic vision were forgotten, only to re-emerge in today's apocalyptic ecological movement. Ironically, the *practical* implications were also part of a continuous line of German agrarian development. As elsewhere in the history of the Third Reich, it is the element of continuity, rather than the discontinuous revolutionary change, that is most striking.

Appendix

MARRIAGE ORDER OF THE SS, 31.12.31, MUNICH,
PROMULGATED BY HEINRICH HIMMLER

1. The SS are a union of Germans of Nordic characteristics, chosen according to distinct aspects.
2. In accordance with the National Socialist conception of things, aware that the future of our people rests on selection, and of the hereditary conservation of racially healthy and sound blood, I hereby introduce the 'marriage approval' for all the unmarried members of the SS, to take effect fully after 1st January 1932.
3. The family, hereditarily healthy, and valuable on account of its German character (in the Nordic sense) is the envisaged purpose.
4. Marriage approval will be given or refused solely on grounds of race or of hereditary health.
5. All SS men who intend to marry are obliged to seek the approval of the Reichsführer of the SS.
6. SS members who marry despite their being refused permission are debarred; they can resign.
7. The decisions on marriage will be made by the racial office of the SS.
8. The racial office will hold the 'SS family records'. The family of SS members will be recorded there after notification of marriage or official acceptance of the request to be recorded.
9. The RF SS, the Director of the Racial Office and the relevant officials of that office are bound by the code of professional confidence.
10. The SS are aware of making an important step forwards with this present order. Jokes, sarcasms and misunderstandings do not affect us; the future belongs to us.

signed HH, RF SS

Notes on Sources

Much of the Ministry of Agriculture's file material was destroyed by bombing during the Second World War, and Darré and Backe both lost the bulk of their correspondence in the same way. Luckily, many of the issues and problems can be reconstructed by using other government material, and sifting through the papers of the Finance Ministry, the Reichschancellery, and the Hitler Adjutant files, etc., for agricultural material. Darré's personal papers are divided between the Goslar City Archives, which has the pre-1933 section, and the Federal Archives at Coblenz, while his letters to his first wife, Alma, are at the Institute of Contemporary History, Munich,. They are, however, restricted, as are Backe's papers at the Federal Archive, Coblenz. Backe's letters to his wife and his documents (passport, references, etc.,) are on microfilm there, and his personal papers, while not copious, contain much that is of interest to the intellectual historian.

One source which presented some difficulty was Darré's diary, which consists of typed extracts taken from his original diary by two friends, after the original was bought back from the Federal Archive, Coblenz by his second wife. She felt that it showed Darré in too bad a light. The original diary has been read by several historians and archivists, although it was unfortunately burnt in the late 1960s. After discussions with the two editors, and with the chief archivist at Goslar, where it can be read but not photo-copied, it seems clear that the typed version is perfectly valid. Its contents are often borne out by material in other private and governmental files, and the material left in does not seem to have been selected with a view to presenting Darré in a specially favourable light; for example, the attack on Hore-Belisha in 1939, and some positive remarks about Hitler. The editors' claim that only material of a personal, prolix and libellous nature has been deleted can, I think, be accepted. One can only regret that so many of Darré's scurrilous but sharp comments have gone for ever. However, I have tried not to base any one aspect of my interpretation on this document, and, for the same reason, have, wherever possible, avoided using defence documents presented at the Nuremberg Trial as a sole source.

It proved impossible to find a German copy of *Neuadel*, and I worked with the French translation of 1939, *La Race*. Page references to *Neuadel* are therefore to the French edition.

All translations, except that on page 64, are by the author.

List of Abbreviations

1. PARTIES AND ORGANISATIONS

AD — *Agrarpolitischer Apparat* (Agricultural-political organisation)

DAB — *Deutsche Ansiedlungsbank* (German Resettlement Bank)

DAG — *Deutsche Ansiedlungsgesellschaft* (German Resettlement Society)

DNVP — *Deutschnationale Volkspartei* (German National People's Party)

DRV — *Deutscher Reichsverein für Siedlungspflege* (German National Association for Settlement Welfare)

DSB — *Deutsche Siedlungsbank* (German Settlement Bank)

DVFP — *Deutsche Völkische Freiheitspartei* (German People's Freedom Party)

EHG — *Erbhofgesetz* (Hereditary Farm Law)

FKG — *Fidei-Kommissgesetz* (Entailed Estates Law)

Gestapo — *Geheime Staatspolizei* (Secret State Police)

G-G — *General Gouvernement* (Non-incorporated Poland 1939–45)

KBF — *Kreisbauernführer* (County Peasant Leader)

LBF — *Landesbauernführer* (State Peasant Leader)

NSDAP — *Nationalsozialistische Deutsche Arbeiterpartei* (National Socialist German Worker's Party)

OKW — *Oberkommando der Wehrmacht* (German Army High Command)

PLRB — *Preussische Landesrentenbank* (Prussian Agricultural Credit Bank)

RfA — *Reichsamt für Agrarpolitik* (National Office for Agricultural Policy)

RFM — *Reichsfinanzministerium* (National Finance Ministry)

RJM — *Reichsjustizministerium* (National Justice Ministry)

RKFDV — *Reichskommissariat für die Festigung Deutschen Volkstums* (National Commissariat for the Strengthening of Germanness)

RMEL — *Reichsministerium für Ernährung und Landwirtschaft* (National Ministry of Food and Agriculture)

RNS	*Reichsnährstand* (National Food Estate)
RuS	*Rasse-und Siedlungsamt* (Race and Settlement Office)
RuSHA	*SS Rasse-und Siedlungs Hauptamt* (SS Race and Settlement Main Office)
SA	*Sturmabteilung* (Storm Division)
SPD	*Sozialdemokratische Partei Deutschlands* (German Social Democratic Party)
SS	*Schutzstaffel* (Protection Squad)

2. ARCHIVAL ABBREVIATIONS

BA	Bundesarchiv Koblenz (Federal Archives, Coblenz)
ED	Darré Nachlass (Institute for Contemporary History, Munich)
G.B.,	*Grosser Bericht*
IfCH	Institute for Contemporary History, Munich
NDG	Nachlass Darré, Stadtarchiv, Goslar (Darré's papers, Goslar City Archive)
TWCN	*Trials of the War Criminals at Nuremberg* (Washington 1949)
UB/TB	Ursula Backe, Tagebuch (Frau Backe's Diary)

3. JOURNAL ABBREVIATIONS

AHR	*American Historical Review*
BuL	*Berichte über Landwirtschaft*
DAP	*Deutsche Agrarpolitik*
DE	*Deutschlands Erneuerung*
DLA	*Die Leichte Artillerie*
HZ	*Historische Zeitschrift*
JBD	*Jahrbuch des baltischen Deutschtums*
JCH	*Journal of Contemporary History*
JHI	*Journal of the History of Ideas*
JMH	*Journal of Modern History*
PSQ	*Political Science Quarterly*
RGB	*Reichsgesetzblatt*
VB	*Völkischer Beobachter*
VjhfZg	*Vierteljahreshefte für Zeitgeschichte*
ZAGAS	*Zeitschrift für Agrargeschichte und Agrarsoziologie*
ZWS	*Zeitschrift für Wirtschaft-und Sozialgeschichte*

Notes

1. See A.C. Bramwell, 'National Socialist Agrarian Theory and Practice, with Special Reference to Darré and the Settlement Movement', Oxford D. Phil Thesis (1982), chapter V.
2. See, e.g., contemporary accounts of Nazi Germany by H. Knickerbocker, *Germany: Fascist or Soviet?* (London, 1932); H. Greenwood, *The German Revolution* (London, 1934); and the valuable economic study by C. Guillebaud, *The Economic Recovery of Germany 1933–38* (London, 1939), which devotes considerable space to agriculture.
3. K. Bergmann, *Agrarromantik und Grosstadtfeindschaft* (Meisenheim/Glan, 1970) pp. 361–4; K. Bracher, *Die deutsche Diktatur* (Berlin, 1969), pp. 168 ff., D. Gasman, *The Scientific Origins of National Socialism: Social Darwinism, Ernst Haeckel and the German Monist League* (New York, 1971), p. xxii; D. Orlow, *The History of the Nazi Party, 1919–1933* (Pittsburgh, 1969), p. 80. Bergmann, *Agraromantik*, pp. 247 ff., dismisses the *Artamanen* settlement group as 'a romantic–illusory irrational and destructive ideology'. F. Grundmann's otherwise fine and comprehensive study of the Hereditary Farm Law is also marred by constant references to 'agrarian fanatics', e.g., *Agrarpolitik im Dritten Reich; Anspruch und Wirklichkeit des Reichserbhofgesetzes* (Hamburg, 1979), pp. 21–8, 151, 157, 228.
4. G. Mann, 'The Second German Empire: the Reich that Never Was', in *Upheaval and Continuity* (London, 1973), pp. 39, 45–6.
5. The autonomy which has been returned to the peasantry as a class in recent English writing, has still not been restored to the Junkers, see D. Blackbourn, *Class, Religion and Local Politics in Wilhelmine Germany* (Yale, 1980), and I. Farr, 'Populism in the Countryside: the Peasant Leagues in Bavaria in the 1890s', in, ed, R. Evans, *Society and Politics in Wilhelmine Germany* (London, 1978).
6. See, for example, the interesting essays in the October 1984 issue of the *JCH* by Milan Hauner, Jeffrey Herf and Jost Hermand.

7. For example, there is the case of the 'internal émigrés', and those with more unambiguous loyalties, such as Gottfried Benn and Ernst Jünger and Martin Heidegger. Where artists and intellectuals began their work before National Socialism came to power, continued to work, fell silent, then recommenced after the war with no noticeable difference in style and subject matter, how is one to distinguish between German and National Socialist? That is not to argue that the distinction is impossible, only that the question is complicated.

8. e.g., K. Ludwig, *Technik und Ingenieure im Dritten Reich* (Düsseldorf, 1974), D. Warriner, *The Economics of Peasant Farming* (London, 1939, repr. 1964), A. Mayhew, *Rural Settlement and Farming in Germany* (London, 1973), M. Tracy, *Agriculture in Western Europe since 1880* (London, 1974); H. Haushofer, *Vom Ersten Weltkrieg bis zur Gegenwart*, vol. 2 of *Ideengeschichte der Agrarwirtschaft und Agrarpolitik im deutschen Sprachgebiet* (Bonn, 1958), cited hereafter as *Ideengeschichte*, K. Brandt, *The Management of Agriculture in the German-Occupied and Other Areas of Fortress Europe: A Study in Military Government* (Stanford, 1953) and the works of Alan Milward. This is obviously not a comprehensive list, nor is a full historiographic treatment appropriate here. The one straightforward, economic and social history treatment of Nazi agricultural policy that has been written, John Farquharson's *The Plough and the Swastika* (London, 1976), which broke new ground and is an excellent survey of the subject, was attacked on publication for its lack of 'an effectively analytical framework', presumably because it contained insufficient adjectives, and is unfortunately out of print.

9. See J. Robinson, *An Essay on Marxist Economics* (London, 1942) and on Marx's anti-semitism, for example L. Feuer, 'Karl Marx and the Promethean complex', *Encounter* (December, 1968), p. 26.

10. F. Engels, 'End of Classical German Philosophy', in ed. L. Feuer, *Marx and Engels, Basic Writings* (London, 1976), p. 270.

11. *Das Bauerntum als Lebensquell der nordischen Rasse* (Munich, 1929) and *Neuadel aus Blut und Boden* (Munich, 1930). Neither book is available in English. There are some extracts from Darré's speeches in B.M. Lane and L. Rupp's *Nazi Ideology Before 1933; a Documentation* (Manchester, 1978).

12. Silvio Gesell (1862–1930) was also Finance Minister of the 1919 Munich Soviet Republic, but see A. Tyrell, 'Gottfried Feder and the NSDAP', in ed. P. Stachura, the *Shaping of The Nazi State*

(London, 1978), pp. 59, 61, who argues that Feder disliked Gesell's 'free money' views.

13. D. Welch, *Propaganda and the German Cinema* (Oxford, 1983).
14. For example, see 'Problems of an Overcrowded Profession', by John Cherrington, *Financial Times*, 15.1.85, on, *i.a.*, French restrictions on farmland tenure, and General Dayan: 'Living on the land is the most essential thing for the individual and the nation ... Jews should stay on the land.' in *Newsnight*, BBC 2, 16.10.81.
15. G. Lewis, 'The Peasantry, Rural Change and Conservative Agrarianism. Lower Austria and the Turn of the Century', *Past and Present*, 80 (1978), p. 142.
16. For Chayanov and von Thünen, see *The Theory of Peasant Economy*, ed. D. Thorner, et. al. (Illinois, 1966).
17. e.g., G.K. Chesterton's introduction to W. Cobbett, *Cottage Economy* (London, 1926), H.J. Massingham, ed., *The Small Farmer* (London, 1947), *passim*, and Emile de Lavelaye, 'Land Systems of Belgium and Holland', in *Systems of Land Tenure in Various Countries* (London, 1870), pp. 244–5, 251.
18. See, e.g., A Polonksy, *The Little Dictators* (London, 1975), for an account of the Successor States and their economic problems.
19. Bergmann, *Agrarromantik*, pp. 362–4, and see also A. Schweitzer, *Big Business in the Third Reich* (Bloomington, 1964), who sees the German peasants as a residual, obstructive group. The 'anti-industrialisation reaction' view is so prevalent that it is hardly worth while giving further examples. The reader may like to start compiling his own list.
20. British Intelligence Objectives Sub-Committee Overall Report, no. 6, *Some Agricultural Aspects in Germany during the Period 1939–45* (London, 1948).
21. As defined in Frank and Fritzie Manuel's *Utopian Thought in the Western World* (Oxford, 1979).
22. M. Heidegger, *The End of Philosophy* (London, 1956), pp. 107, 110.
23. Ibid., p. 109.

NOTES TO CHAPTER ONE

1. *Neue deutsche Biographie*, p. 517; *Der deutsche Führerlexicon*, p. 34–5; Reischle, H., *Reichsbauernführer Darré: der Kampfer um Blut um Boden* (Berlin, 1933), hereafter referred to as *Darré*; photographs, Darré Nachlass, City Archive, Goslar (hereafter refered to as NDG) file 21; Darré's curriculum vitae, 31.12.29, Federal Archives (hereafter

referred to as BA), NL94I/9, and R. Oskar Darré, *Meine Erziehung im Elternhause und durch das Leben* (Wiesbaden, not dated, c. 1925).

2. R. Oskar Darré, *Erziehung*.
3. Darré's father and his drinking problem, Darré to Alma (née Stadt), his first wife, 25.11.23., Darré papers, Institute for Contemporary History, Munich, ED 110/8 (all the Institute Darré papers which have the file no ED, will simply be numbered under that headling in these notes); Carmen Darré to Darré, referring to 'Darréan idleness, excluding our father, naturally,' 20.X1.19, and 8.X1.19., NDG/76a.
4. Darré's school papers and reports, NDG/462. *King's College Magazine*, vol. xxxv, 3, 1933, pp. 34–5, and correspondence with KCS Registrar and author, 1978. See Reischle, *Darré*, p. 13, where the effect of his English education was stressed. Darré had deleted any reference to it from his c.v. of 1929.
5. W. W. Schmokel, *Dream of Empire: German Colonialism, 1919–1943* (London, 1964), pp. 44–5; *Meyer's Lexicon* (Leipzig, 1927), p. 1571.
6. See Darré's passes and passports, including Argentinian consular material, NDG/461, Argentinian passport of 1927, and correspondence with the Argentinian consul in Wiesbaden, 15.5.22, NDG/77a, and IfCH Interrogation, 10.3.47, p. 2.
7. Diary, ED 110/6, Reischle, *Darré*, pp. 19–21.
8. Diary, 8.12.18, and date unclear, file pages 21, 27–8, ED 110/6, and note from the Witzenhausen 11 Mar. Brigade, 14.8.19, NDG/76a.
9. In ED 110/6.
10. For Darré's Stahlhelm membership, see letters to family and friends over this period, 31.5.27, Darré still on their membership list and receiving Stahlhelm literature, NDG/83a, and IfCH, Interrogation, p. 6.
11. Homeopathic chemist to Darré, 10.5.22, NDG/77: Nature healer diagnosis, January 1924, NDG/78a (tendency to TB).
12. R. O. Darré, *Erziehung*, p. 32.
13. Ibid., p. 32.
14. Passes and restrictions, see letters Oskar Darré to Darré, 1.8.19, 10.8.19, on foreign affairs, 25.2.20, NDG/76a.
15. Arthur Kracke (ex-soldier friend of Darré's) to Darré, 21.5.29, NDG/76c. F. Carsten, *The Reichswehr and Politics 1918–1933* (Oxford, 1966), pp. 22–3: E. E. Dwinger, *Wir Rufen Deutschland* (Leipzig, 1932), p. 16, quoted in R. Taylor, *Literature and Society in Germany; 1918–1945* (Sussex, 1980), p. 15 and see also 257, where Taylor comments that Dwinger was

a literary spokesman for the so-called conservative revolution of the late 1920s and early 1930s … entitled to be seen, at least in part, as men engaged in the not dishonourable task of salvaging a national self-respect from the failures of an inglorious regime, not as rabble-rousers in the service of National Socialism.

16. V. Berghahn, *Der Stahlhelm; Bund der Frontsoldaten* (Düsseldorf, 1966), p. 7.
17. One interesting biographical account of the *Freikorps* available in English is Friedrich Glombowski, *Frontiers of Terror* (London, 1935), tr. K. Kirkness. See also I. Morrow, *The Peace Settlement in the German–Polish Borderlands* (London, 1936), and E. v. Salomon, *Der Fragebogen* (Hamburg, 1951). See also M. Sering and C. von Dietze, ed., *Agrarverfassung der deutschen Auslandssiedlungen in Osteuropa* (Berlin, 1939).
18. A. Nicholls, *Weimar and the Rise of Hitler* (London, 1968), ch. 5.
19. G. Field, *Evangelist of Race: the Germanic Vision of Houston Stewart Chamberlain* (Columbia, 1981), p. 408.
20. German Colonial School to Oskar Darré, 3.5.19, NDG/76a: Diary, 29.3.19, ED 110/6, and 9.1.19, ED 110/6. The eldest Darrés, Richard Walther and Carmen, were very close, and were to remain friends. They formed an alliance to protect the younger members of the family, especially Ilse, from the father, Carmen even attempted to remove Ilse from an unhappy home atmosphere in 1919, when she complained that her father was 'such a fuss-pot, and blind to the good things in life'. Carmen to Darré, 3.10.19, 8.11.19, NDG/76a.
21. Diary, 14.10.19, ED 110/6.
22. Darré to Alma, 19.10.19, ED 110/6.
23. Darré to Alma, 24.11.20, ED 110/6. Another echo of Bölsche's nature-worship appeared when he wrote, 'You know how much I feel a sense of affirmation towards Nature's love-life (*Liebesleben*). In the same way, I affirm physical love between people.' 20.11.20, ED 110/6. Darré persuaded her to take up gymnastics, writing that the 'body is a reservoir of strength; it should be regarded correctly', 30.10.23, ED 110/8. Ironically, Darré damaged a leg in a 400 metres race in 1936, and his absence from his office at a crucial period gravely weakened his political position.
24. Darré to Alma, 20.11.20, ED 110/6.
25. In late 1920 they sent specimen of each other'a handwriting to a graphologist. Alma's apparently showed that she was strong-willed, truthful, affectionate, impatient. Darré 'had a lively, suffering temperament; self-discipline; a feeling for what is good, noble and honourable.' Stefanie von Tschudi, 13.11.20, ED 110/6.

26. 27.11.20, NDG/80, and see subsequent correspondence, and BA NL94I/9, which contains letters from Darré to his Hanover lawyers, and the lawyers' letters to the new headmaster at Witzenhausen.
27. Darré was nominated for membership of the officer's union, the von Scharnhorst, 14.8.29, NDG/85a. For details of degree marks, see NDG/462.
28. See letters NDG/77a and 77b, e.g., 5.5.22, 23.2.22, 1.8.19, 1.8.19, and the reference to Darré's heart problems (somewhat sarcastic), Oskar Darré to Darré, 21.8.19, NDG/76a.
29. Darré insisted on taking out an individual membership in the *Vereinigung deutscher Schweinezuchter*, although the farm already had a joint membership, 29.XI.21, NDG/77b. The letter refusing him the post, 18.1.22; correspondence concerning smallholdings and farms, November 1921 to March 1922, all NDG/77a.
30. Darré's aunt, 8.12.22, NDG/77b; GCS to Oskar Darré, 16.11.78, NDG/76c; Oskar Darré to Darré, 10.8.19, NDG/76a, and postcard, Ilse to Darré, no date, NDG/76b.
31. Darré's debts, see 16.4.22, NDG/80, and subsequent correspondence. Erich Darré to Darré, 25.7.22, NDG/77b. Erich Darré shared the family's Anglophilia. He sent Darré a card from one visit to England, written in fractured English; 'the beer causes that we all double see'.
32. See Oskar Darré to Darré, 1920–22, NDG/76c and especially 23.5.20 and 28.5.20.
33. Darré to Oskar Darré on Lagarde, February, 1925, NDG/80. *Tonio Kröger* was among the books forwarded to him from Gut Aumühle, 16.8.22, NDG/80. For the von der Vogelweide quote, Darré to Oskar Darré, 11.11.23, NDG/80. Darré to Alma, c. October 1923 (file pages 881–2), ED 110/8, and 21.2.22, NDG/77a. One letter to his wife refers to Kotzde's autobiography. Kotzde was the leader of a German youth group, *Adler and Falken*, and co-founder of the *Artamanen*, see G. Mosse, *The Crisis of German Ideology* (London, 1964), p. 116, and 8.12.16, ED 110/9.
34. Halle curriculum, NDG/80. Reports, NDG/462. See also Darré's letters to his wife concerning a study of the hereditary transmission of musical ability carried out by Halle students for Professor Krämer, 12.2.23, ED 110/8.
35. 'Internal Colonisation', *Deutschlands Erneuerung*, 1926. Cf. Schmokel, *Dreams of Empire*, p. 91.
36. South American friend to Darré, 15.4.25, NDG/79, and 10.5.27, NDG/83b.

37. Darré to Oskar Darré, 19.7.25, on the Prussian Ministry of Agriculture's refusal to recognise Witzenhausen's traning, NDG/80, and Ilse to Darré, 11.6.27, NDG/83a.
38. Report, NDG/78a. Dancing lessons, Oskar Darré to Darré, 21.8.19, NDG/76a.
39. See passports in NDG/462.
40. NDG/462. Darré was involved in a serious quarrel at Giessen, that ended up with the student Court of Honour there. See correspondence, February and March, 1926, NDG/81a.
41. See bibliography, BA NL 94II/49.
42. See G. Stokes, 'The Evolution of Hitler's Ideas on Foreign Policy', in P. Stachura, ed., *The Shaping of the Nazi State* (London, 1978). I am indebted to Dr Eleaner Breuning, of University College, Swansea, for information about the Prussian Ministries and their representatives abroad.
43. Darré to Dr Kurschat, 29.7.26, Darré to the head of the East Prussian Trakhener Society, 6.8.26, both NDG/81b.
44. This quality was to stand horses and owners in good stead when the remarkable operation to rescue the bloodstock from the Russians took place in 1945. Barely half the horses survived the flight across the frozen Haff. The East Prussian Chamber of Agriculture summoned Darré to a meeting with the Russian Consul, 15.3.27, NDG/82. Requests for posts for relatives, 5.6.28, NDG/83b and subsequent corr. For the Trakheners, see Dr Fritz Schilke, *Trakhener Horses; Then and Now* (Oklahoma, 1977), and for the rescue attempt, pp. 116–7.
45. Darré to Alma, undated, but November 1926 from the contents, ED 110/9.
46. Darré to Alma, 15.2.27, ED 110/9. Darré's report on Finland, NDG/400.
47. Metzger's report, NDG/81a. Metzger knew Dr Otto Auhagen, agricultural economist, whose thesis had concerned the superior qualities of small farm productivity. Auhagen visited Russia several times during the 1920s. In 1939, in his seventies, he was apparently drafted into the SS, to help draw up plans to settle German peasants in the East. Darré's plea to Metzger, 5.5.27, and Metzger to Darré, 10.5.27, both NDG/83a. Darré's acquaintances at this time included several East Prussian landowners, such as Rodbertus-Januschau, vom Zitzewitz, Freiherr von Wangenheim, of the Agrarian League, and von Schwerin, who advised him to contact Baron Manteuffel, 15.9.26, NDG/82.

48. Edgar Jung to Darré, 11.5.28, NDG/84a. See also chapter Three. For Von Behr to Darré, 14.4.27, NDG/83a.
49. Reischle, *Darré*, pp 29–30. Darré to Alma, undated letter, ED 110/9. Telegram, NDG/85a and Darré to Alma, 21.2.29, and 24.2.29 ED 110/9.
50. Darré to Jacob Stadt, his father-in-law, 1.6.29, ED 110/9. IfCH, Interrogation, 10.3.47, p. 4. The court case is mentioned in a letter from Darré to Duckwitz, Legation Councillor at Reval, 31.7.29, NDG/89a. Sohn-Rethel. *Economy and Class Structure of German Fascism* (London, 1978), p. 69. Sohn-Rethel thought that Reusch, head of a huge industrial combine, was able to find Darré a post in the National Ministry of Agriculture, but offers no reason why Reusch should have been able to create and dispose of civil service posts at will, nor why he should have favoured Darré, although Sohn-Rethel claims that Darré was so exceptionally stupid that he was considered their ideal protector by German industry. Sohn-Rethel is an excellent example of the cavalier attitude to facts, and the easy conspiratorial theorising which might be called the 'Claud Cockburn syndrome'.
51. Letter of rejection, *DE* to Darré, 23.6.25, NDG/79. The essay is in NDG/80. See Darré to Alma, 19.1.23, ED 110/8.
52. For a picture of the demonstration, see *Der Stahlhelm; Errinerungen und Bilder* (Berlin, 1932), p. 185. Darré wrote to his father from Giessen that 'in the summer of 1925 it was risky to wear the Stahlhelm insignia in public; especially at the time of the Rathenau murder. Three years later, the Stahlhelm rules the streets in Halle.' 13.12.25, NDG/80. Darré attended these rallies and veteran's reunions (banned, but held under the guise of a 'Prussian Day') between 1922 and 1925, see e.g. 25.X1.22, NDG/77b.
53. Darré to Oskar Darré, 13.12.25, NDG/80.
54. Darré to Oskar Darré, 11.11.23 and 19.11.23, NDG/80.
55. For the DVFP, see J. Noakes, *The Nazi party in Lower Saxony* (Oxford, 1971), pp. 28–9. The 1937 attack is by Fr. von Wangenheim, in 'Dietrich Eckhart und Ich', *Der deutsche Bauer*, 15.3.37. See Darré to Alma, 12.2.23, ED 110/8, and see chapter Four for Alma Darré's Party membership.
56. J. M. Keynes, *The Economic Consequences of the Peace* (London, 1920), p. 234, quoting a report published in Sweden in April 1919.
57. N. Hamilton, *The Brothers Mann* (London, 1978), p. 216, and cf. P. Pulzer, *The Rise of Political Anti-Semitism in Germany and Austria* (London, 1964), p. 5, '[Jews] contributed to [liberalism's]

establishment, benefited from its institutions, and were under fire when it was attacked'.

58. Darré to Alma, anti-semitism, 19.1.23, the 'Soviet star', 21.7.23, ED 110/8. For student anti-semitism, see Mosse, *Crisis*, pp. 268–9.

NOTES TO CHAPTER TWO

1. G. G. Field, 'Nordic Racism', *Journal of the History of Ideas*, 38 (1977), 523, and introduction to *Northern Legends* (London, 1902), tr. M. Bentinck, on 'Northern', 'Teutonic' and 'Deutsch'.

2. H. S. Chamberlain, *Foundations of the Twentieth Century* (London, 1911), tr. John Lees, vol. 1, pp. xvi, 522, 529, and H. F. K. Günther, *Kleine Rassenkunde des deutschen Volkes* (Munich, 1922). Günther was appointed Professor of Racial Hygiene at Jena in 1931 by Frick, then National Socialist Minister for Education in Thuringia. The literary journal *Die Zone*, 24.12.33 (Paris and Vienna), ed. Emil Szittya, carried a picture story showing various important German political figures, including Ludendorff, von Epp and Goering, with the heading 'Racial Beauty does not exist in Germany'.

Anyone could play the game of racial categorisation. One writer who criticised H. S. Chamberlain's work for a lack of scholarliness attributed it to a 'powerful Anglo-Celtic inclination', Field, *Evangelist of Race*, p. 234.

3. C. S. Lewis, *Surprised by Joy*, (London, 1955), p. 74, and see also pp. 75, 78, 123. Two of his friends became Anthroposophists, pp. 194, 196–7.

4. Field, 'Nordic Racism', p. 533.

5. For the *Bauernhochschulbewegung* see K. Bergmann, *Agrarromantik* and Field, 'Nordic Racism', p. 532. Lenz is quoted in H. Lutzhöft, *Der nordische Gedanke in Deutschland 1920–1940* (Stuttgart, 1971), p. 136. For *Varuna* see Mosse, *Crisis*, p. 115, Lutzhöft, op. cit., pp. 134–6, Field, op. cit., p. 529.

6. e.g. M. Weber, 'Power', in *Essays in Sociology*, ed. H. Gerth and C. Wright Mills (London, 1967), pp. 177–8, and see H. F. K. Günther, *The Racial Elements in European History* (London, 1927), *passim*.

7. See Field, *Evangelist of Race*, pp. 296–7, and Chamberlain, *Foundations*, pp. 517–20.

8. See description in Lutzhöft, *Nordischer Gedanke*, *passim*, Field, 'Nordic Racism', p. 532, and Knut Hamsun's novels, especially *Mysteries*, *Growth of the Soil* and *Chapter the Last*.

9. e.g. Lehmann to Darré, correspondence in NDG/84a: Konopacki-

Konopath to Darré, 19.1.27, NDG/84a. Günther asked Darré to push his book in England if he had connections there, as 'the press is Jewish influenced or Jewish led', 29.XI.27, NDG/84a.

10. For the ban on Lanz von Liebenfels, see E. Howe, *Astrology in the Third Reich* (London, 1984), p. 111. Schultze-Naumburg and secrecy, see his letter to Darré, 30.7.30, in NDG/87a, with the minutes of 'a secret meeting at Saaleck' and talk of *hochwertigen Rassenkerns*. See also Darré and his Friends of the German Peasantry, 1939, BA NL 94II/1d, and Günther on Protestants preferable to Catholics, *Racial Elements*, p. 219.

11. See R. Lougee's *Paul de Lagarde* (Cambridge: Mass., 1962), for an excellent account of Langbehn (and Lagarde), and *passim* in Pulzer, *Rise of Political Anti-Semitism*.

12. E. von Salomon, *Die Fragebogen* (Hamburg, 1951), pp. 30–31.

13. Anonymous critic, describing himself as a 'Lower Saxon Peasant', *Est, Est, Est: Randbemerkungen zu 'Rembrandt als Erzieher'*, n.d., p. 22. Darré to Alma Darré on Langbehn, 24. XI.26, ED 110/9.

14. Darré to Alma Darré, 14.1.24, ED 110/8, and 9.10.26 and 5.12.26, recommending Günther's *Style and Race*, ED 110/9.

15. *Neuadel*, quotations here taken from the French translation, *La Race*, 1939, p. 114. Darré to Alma Darré, 14.1.24, ED 110/8. Darré on Prussia, 21.XI.25, ED 110/8, 'The meaning of Prussianness is not just a kingdom, but the pure concept of Duty ... heterogeneous elements formed a unity through the Prussian Idea'.

16. R. W. Darré, 'Walther Rathenau und das Problem des nordischen Menschen', *DE*, vii (1926). Salomon, *Die Fragebogen* (English translation, *The Answers of Ernst von Salomon*), p. 57. Rathenau and Theodore Roosevelt both liked Houston Stewart Chamberlain's work, though Roosevelt criticised its 'startling inaccuracies', Field, *Evangelist of Race*, p. 466.

17. Coudenhove-Kalergi, *Neuadel* (Leipzig, 1922). As a guru of the European movement, Kalergi has been the subject of at least one hagiographic biography by Euro-hacks. The pamphlet received little attention.

18. Dr Oskar Levy, introduction to Gobineau, *The Renaissance*, tr. P. Cohn (London, 1913), and see R. Hinton-Thomas, *Nietzsche in German Politics and Society, 1890–1918* (Manchester 1984), pp. 101–2.

19. R. W. Darré, 'Walther Rathenau und die Bedeutung der Rasse in der Weltgeschichte', *DE*, I, 1928.

20. For correspondence concerning the Nordic Ring, membership lists and meetings, see NDG/84a. For *Die Sonne*, see Mosse, *Crisis*, pp.

78, 83, 106. Darré refers to his first meeting with Günther, through the 'Rathenau' article, 30.1.32, ED 110/10.

21. For Vietinghoff-Scheel, see Pulzer, *The Rise of Political Anti-Semitism*, p. 305.
22. Marie Adelheid Princessin Reuss to Darré, 12.1.26, and Darré's reply from Insterburg, not dated, both NDG/84a. Hanno Konopacki-Konopath to Darré, 19.1.27, NDG/84a.
23. cf. here, Jeremy Noakes, 'Nazism and Eugenics: the Background to the Nazi Sterilisation Law of 14 July 1933', in *Ideas into Politics* (London, 1984), ed. R. Buller, H. Pogge von Strandmann, and A. Polonsky, p. 75: 'For both the eugenics programme to improve German racial hygiene and the programme of anti-semitism ultimately derived from a common perspective, a perspective which viewed man and society not simply in biological terms but from a particular Social Darwinist viewpoint'. As mentioned in the Introduction to this book, work can be scrupulously scholarly, and carefully annotated in its factual detail, yet be misleading in its general conclusions. Incidentally, the term Social Darwinism first came into common use in the 1940s, after Richard Hofstadter's influential book on the subject, *Social Darwinism in American Thought* (New York, 1944).
24. Details of conference, NDG/87b and see also Nordic Conference Programme for 1930, in ED 110/11, with list of participants, such as the *Kampfbund* for German Culture, the German Richard Wagner Society, etc.
25. Professor V. Stranders, MA; he was a member of the Territorial Army up to January 1915, and at some time during the First World War was a captain in the Royal Air Corps. He went to Germany with the Allied Control Commission as head of translations for the Air Branch in 1919. See his two books: *Die Wirtschaftsspionage der Entente* (Berlin, 1929) and *Vernichtung über Deutschland* (Berlin, 1933), both extremely anti-French.
26. Mosse, *Crisis*, pp. 71–2. Darré to Hitler, 11.12.39, BA NS 10/37, enclosing von Leers' comments on Hore-Belisha. Union of German Druids to Darré, 19.10.34, BA R16/2141.
27. Postcard, Darré to Konopath, NDG/84a.
28. Günther to Darré, suggesting that he answer Kern, 5.11.27, NDG/84a.
29. For nomadic architecture, see my chapter Four, and B. Miller Lane, *Architecture and Politics in Germany: 1918–1945* (Cambridge, Mass., 1968), p. 159. Konopath told Darré that through Kern's

arguments, 'We have surrendered our position to the Jews', 17.4.28, NDG/84a.

30. On Faustian and nomadic nature of the Nordics, Günther to Darré, 5.11.27, NDG/84a, and on Darré's understanding of racial questions, Günther to Darré, letter not dated, NDG/87a.

31. Lehmann to Darré, stressing the need for the sterilisation of the hereditary sick, especially blind and deaf people, 28.7.28.: Darré too outspoken on sexual matters, 30.7.28., both NDG/84b. For a discussion of Lehmann, see G. Stark, *Entrepreneurs of Ideology: Neo-Conservative Publishers in Germany, 1890–1933* (Chapel Hill, 1981).

32. Günther to Darré on his style, undated letter, NDG/84a.

33. R Eichenauer, 'Heraus mit Eurem Flederwisch', *Die Sonne* (Feb. 1931). Eichenauer to Darré, 29.12.28, and subsequent correspondence, all NDG/84b.

34. B. Miller Lane and L. Rupp, *Nazi Ideology Before 1933* (Manchester, 1978), p. xxvii. Darré's book eventually appeared as *Das Bauerntum als Lebensquell der nordischen Rasse* (Munich, 1929), cited hereafter as *Das Bauerntum*. For the request to delete foreign words, see Günther to Darré, 5.X1.27, NDG/84a, and Lehmann to Darré, 14.X1.28, NDG/84b. Darré earned 1350 marks for the book. It eventually sold some 40,000 copies, but in the first year only sold about 750.

35. For Darré the unwordly philosopher, see H. Schacht. *My First Seventy-six Years* (London, 1955) p. 328, and Kerrl; also my chapter Five, and Clifford R. Lovin's fine thesis on Darré and his programme, drawn from secondary sources only, and thus a particularly impressive achievement, stressing Darré's role as practical agrarian economist, 'German Agricultural Policy, 1933–36,' Ph.D. Thesis, University of North Carolina, 1965.

NOTES TO CHAPTER THREE

1. K. Meyer, 'Lebensbericht', unpublished memoirs, pp. 96–7. *The Goebbels Diaries* (London, 1948), tr. and ed. Louis Lochner, p. 168. Lochner's definition of 'blood and soil' includes the explanation: 'Nordic blood was the best in the world, and there could be nothing more perfect than to have been born on German soil', also p. 168.

2. E.g. D. Welch, *Propaganda and the German Cinema, 1933–45* (Oxford, 1983), pp. 96, 137–8. I. Berlin, 'Benjamin Disraeli and Karl Marx', in *Against The Current* (Oxford, 1982), p. 268.

3. Interview with Dr H. Merkel and interview with F. Krausse, April, 1984.
4. See pictures in B. Hinze, *Art in the Third Reich* (Oxford, 1980), esp. pp. 86–7, 91–7, and *Deutsche Agrarpolitik*, as *Odal* was renamed.
5. See 'The Folk-Community', *Rising*, 2 (1982), p. 4.
6. Damaschke pops in and out of the literature according to whether he is seen as a suspect proto-Nazi (repulsive but interesting) or as a kind of German Fabian (nice but boring). His autobiography, *Aus Meinem Leben* (Leipzig, 1924), is frankly dull. The late nineteenth century saw a surge in socialist, Utopian, teetotal social-reformers, and Damaschke belongs in that galère. On the population decline, see N. Jasny, *Bevölkerungsrückgang und Landwirtschaft* (Berlin, 1931), p. 8.
7. See e.g. the Spanish nationalist writer Menendez y Pelayo, who also liked to contrast Mediterranean traditions with Northern European ones; D. Foard, 'The Spanish Fichte', *JCH*, 14 (Jan, 1979), esp. p. 91. Writers since Gibbon have been puzzling over the causes of political and cultural distinctions between North and South Europe.
8. R. W. Darré, 'Stedingen', *Odal*, 3 (934–5), and see also R. Cecil, *The Myth of the Master Race; Alfred Rosenberg and Nazi Ideology* (London, 1972), pp. 96, 165–6.
9. R. W. Darré, 'Innere Kolonisation', *DE* (1926), pp. 154–5.
10. R. W. Darré, 'Zur Wiedergeburt des Bauerntums', *DE* (1931), reprinted *Blut und Boden* (Munich, 1941), p. 60.
11. Hugenberg and Darré, see ed. J. Noakes and G. Pridham, *Documents on Naziism, 1919–1945* (London, 1974), p. 386; Charles W. Smith, Introduction to G. Ruhland, *The Ruin of the World's Agriculture and Trade* (London, 1896). The 16.1.33. *Milchgesetz*, introducing a Price Commissioner, is in BA R43II/192.
12. R. Bertrand, 'Le Nationale-Socialisme et La Terre', *Études économiques* (1938), esp. p. 27, is a fairly sympathetic account:

> Thus reduced to a technique of intervention, German corporatism appears heavy with chains and light on spiritual élan, compared with ideal corporate construction. But is this through the will of the masters, or by inevitable conditions?

See also introduction to the French edition of *Neuadel, La Race* (Paris, 1939). Darré was still paying his debt to Ruhland in 1941;

see Dr H. Reischle, 'Von Ruhland zu Darré', *Die Landware* (Berlin, 14.9.41), p. 1. See Rundschreiben no. 65, NDG/142, and articles in *Odal* on Ruhland, April and November 1935.

13. Telegram from Verband der Mitteldeutschen Industrie to von Schleicher, 13.1.33, BA R43II/192.
14. B. Miller Lane, *Architecture and Politics in Germany 1918–1945* (Cambridge, Mass., 1968), pp. 136–7, and see Schultze-Naumburg, 'Müssen wir künftig in asiatischen hausern wohnen?', 12.3.31, article in Darré's files, ED 110/11. The President of the Agricultural Chamber wrote to von Schleicher complaining about architects in agriculture, 7.1.33, BA R43II/192.
15. Schultze-Naumburg, in report to the Friends of the German Peasantry Society, BA NL94II/1d; Miller Lane, *Architecture and Politics*, p. 133; for Mies van der Rohe, see D. Watkins, *Morality and Architecture* (London, 1977), p. 97.
16. Haeckel, *The Wonders of Life* (London, 1905), p. 157.
17. Lenz, quoted in Loren R. Graham, 'Science and Values; the Eugenics Movement in Germany and Russia in the 1920s', *AHR*, 82 (1977), p. 1143, and epigraph to Coudenhove-Kalergi's *Adel* (Leipzig, 1922).
18. Langbehn, quoted in P. Pulzer, *The Rise of Political Anti-Semitism in Germany and Austria* (London, 1964), p. 240.
19. Haushofer, *Ideengeschichte*, p. 31. Alfred Kelly's excellent analysis of Darwinist thought in Germany, *The Descent of Darwin; the Popularisation of Darwinism in Germany, 1860–1914* (North Carolina, 1981), gives a comprehensive and fascinating account of Bölsche. Bölsche went on to write a biography of Haeckel, which, in the heady intellectual climate of pre First World War England, was almost immediately translated into English.
20. Moeller van den Bruck, *The Third Reich* (Berlin,/London, 1923), p. 22.
21. Quoted in Cecil, *The Myth of the Master Race*, pp. 119–120.
22. Darré's diary, 20.9.33.
23. Darré's diary, 2.6.34.
24. See Darré's circular to members of the National Peasant Council, 10.3.36, BA R16/2045.
25. Rosenberg to Darré, 7.1.26, BA NS2/41, and Darré's circular to the National Peasant Council, 10.3.36, BA R16/2045. For the Nordic Gathering, see circular to N.P.C., 23.5.38, BA NS26/947. See Lutzhöft, *Nordischer Gedanke*, p. 332, where Lutzhöft charts the Nordic movement's decline and resentment at its loss of influence and contacts.

26. Darré, SS affidavit, p. 8.
27. Darré's Diary, 22.6.39.
28. R. W. Darré, 'Blut und Boden', *Odal* 1935).
29. M. Hauner, in *India in Axis Strategy* (Stuttgart, 1981), p. 49, refers to the 'anti-colonial but less important group of "radical agrarians" ... centred around Walter (*sic*) Darré'.
30. The reprint is in *Blut und Boden* (Munich, 1940).
31. 'Farming, Community and State' (Lane and Rupp translation), *VB*, 19–20.3.31. the *DE* version of 1930 is slightly longer. Bergmann, in *Agrargomantik*, p. 308, comments on the omission, but misses the point, that Darré had deliberately passed up an attempt to align himself with the regime, and had instead disassociated himself from the eastward expansion that in the end actually took place.
32. Hitler, *Mein Kampf* (London, 1939), pp. 124–5.
33. Dr Edgar Jung to Darré, 27.10.27, NDG/84a.
34. Darré did quote from Moeller in *Neuadel*, p. 255. Moeller committed suicide in 1925, and was therefore spared the dilemma of other messianic neo-conservatives in 1933. His phrases and rhetoric turn up in Backe's notes around 1931, possibly through Darré's influence. For German eugenics and eugenicists, see Dr Paul Weindling's forthcoming book on the subject. Darré had sympathetic criticism from the *Deutsche Ritterband* in 1929. See NDG/84a and 157 for Darré's fan-mail.
35. See debates in the Prussian *Landtag*, Eheberatungstellen, Drucks. nr. 735 (Reports and Petitions), 1925–8. For Ploetz, see Graham, 'Science and Values', p. 1143, and Dr Paul Weindling, unpublished seminar paper, Oxford, March, 1982, 'Population Policy and Eugenics in Germany, 1900–1930'.
36. For the British syphilis figures, W. MacNeill, *Plagues and Peoples* (London, 1977), p. 285. The marriage certificate, urged by Agnes Bluhm in 1905, was taken up by the Bund für Mutterschutz in 1907. See also Grete Meisel-Hess, 'Mutterschutz als soziale Weltanschauung', *Die Neue Generation*, 'the race with the greatest vitality, the white race, should, as the bearer of the highest culture, spread itself as widely as possible', quoted in R. Hinton-Thomas' valuable new book, *Nietzsche in German Politics and Society, 1890–1918* (Manchester, 1984), p. 93.
37. Dr Weindling, op. cit., and lectures, Michaelmas, Oxford, 1984.
38. Graham, 'Science and Values', p. 1143.
39. G. Sanvoisin, quoted in the *Bulletin des Halles*, quoted in *Action*

Francaise (9.7.39). Lenz is quoted in Lutzhöft, *Nordischer Gedanke*, and see Darré to Lenz, 10.2.32, NDG/87a.

40. E.g., Haase to Backe, BA NL75/10. Other works recommended were by von List, Günther, Rosenberg and Ludendorff.
41. Stark, *Entrepreneurs of Ideology*, p. 195, but cf. Lehmann to Darré over slow selling of *Das Bauerntum* in the first two years, only 750 copies by 1930, 6.2.31, NDG/86b.
42. *Das Bauerntum*, p. 428.
43. *Neuadel*, p. 225, and see also *Das Bauerntum*, pp. 226, 367.
44. *Neuadel*, p. 259.
45. cf. J. Lee, 'Labour in German Industrialisation', p. 459.
46. E.g., L. Stoddard, *The Revolt Against Civilisation* (1922, repr. New Orleans, 1950), and M. Grant, *The Passing of the Great Race* (New York, 1921).
47. 'Eugenics for the German People', op. cit.
48. Konopath, circular to the 'Leader of bündischen Jugend', 6.8.30, NDG/87a.
49. Darré to Konopath, 17.8.30, NDG/87b.
50. R. W. Darré, 'Die Frau im Reichsnährstand', *Odal* (March, 1934).
51. Interview with Frau Backe, January, 1981.
52. *Neuadel*, pp. 216–7.
53. Ibid., pp. 217–8.
54. Darré to Alma, 25.11.23, and see also 21.11.25, 25.11.25, 19.1.23, 21.7.23, all ED 110/8.
55. *Neuadel*, pp. 226, 229, 233.
56. Ibid., pp. 241–2.
57. Grundmann, *Agrarpolitik*, pp. 123–4.
58. Darré to SS RuSHA (horses), 15.4.35, BA NS2/37: list of educational events, BA NS2/108. In 18.5.37, Himmler told Darré that Hitler had asked the SS RuSHA to speed up its processing of SS applications to get married; they were some 20,000 behind. 'He told me, half smiling, half crying, that RuSHA was simply preventing SS marriages', BA NS2/41.
59. R. W. Darré, circular to RNS, 14.10.35, BA R16/2045.
60. *Neuadel*, p. 200.

NOTES TO CHAPTER FOUR
1. For Habicht, see R. L. Koehl, *The Black Corps* (Wisconsin, 1983), p. 105. The correspondence with Schultze-Naumburg is in NDG/84a. For the reference to Günther see Darré to Alma, 8.1.29, ED 110/10. See also Holfelder to Hitler, 6.6.29, BA NS26/1285.

2. See NDG/84a. For Jünger and Heidegger, copy letter in author's possession, Ziegler to Darré, 5.11.52. Johst was given honorary membership of the National Peasants' Council in 1936, 1.9.36, BA NS26/947.

3. See Dr Horst Rechenbach to Darré, 14.6.32, NDG/87a. Rechenbach's old teacher, a Professor of the Ethnology-Anthropological Institute at Leipzig, sent Darré best wishes for his 'extremely interesting and necessary work', 13.9.30, NDG/87a.

4. Farquharson, *Plough and Swastika*, pp. 16–17, citing Professor Horst Gies' 'NSDAP und Landwirtschaftliche Organisation', *VjhZg* (1967), p. 343. Darré on Hugenberg to Lehmann, 15.12.29 and Lehmann to Darré, 5.11.29, both NDG/437a.

5. Lane and Rupp, in *Nazi Ideology*, p. xxi, suggest Himmler and perhaps Strasser. Friedrich Grundmann, *Agrarpolitik*, suggests Werner Willikens, and the fact that a typed copy of the manifesto is in Backe's papers (he was a close friend of Willikens) hints that there is something in the idea. But according to Erwin Metzner, 'Keeper of the Seal' for the National Peasants' Council, it was written by Himmler and Hierl (see letter in author's possession). There seems no evidence for the idea that it was written by Darré or Kenstler, and the prevalant text-book suggestion that it was Darré's work shows the cavalier attitude to basic historical research by too many writers on this subject.

6. 'Word for word', see Darré to Ziegler, 27.3.30, NDG/87a.

7. The memo is in BA NS35/1.

8. Darré to Kenstler, 12.4.30, NDG/94.

9. Darré to Ziegler, 7.5.30, NDG/94. The Institute was to be in Weimar, but financed from the NSDAP headquarters in Munich, see Darré to Alma, 23.7.30 and 27.3.30, ED 110/10. Darré to Alma regarding Schultze-Naumburg's card from Frau Bruckmann, 18.4.30, ED 110/10.

10. Darré to Kenstler, 25.4.30, NDG/94.

11. See Darré's Interrogation by Dr Werner Klatt, 1945, Wiener Library, N47 Darré J 36, p. 1. Darré to Lehmann asking for a loan, 28.5.30 and Darré to Kenstler re; money, 25.5.30, NDG/94.

12. See draft letter in NDG/94, and Darré to Kenstler, 28.5.30, ibid, which mentions the phone call.

13. See Darré to Pietzsch discussing the agricultural crisis, 18.2.32, NDG/87a.

14. Stark, *Entrepreneurs of Ideology*, p. 228.

15. Conference, Darré to Alma, 13.6.30, ED 110/10.

16. Konopath, 25.2.31, and quotation, Darré to Konopath, 18.8.30, both NDG/87a. For Darré being too busy to have time for the Nordic Ring, see 14.11.30, and 'fighting to secure a place for Blood and Soil ideals in the Party', 24.4.31, both NDG/87a.
17. Princess zur Lippe to author, 8.12.80.
18. Darré to Alma, 8.7.30 and 27.9.30, both ED 110/10.
19. 30.9.30, NDG/87a.
20. Darré to Alma, before a visit to Weimar with Hitler, 11.4.31, and quotation, 12.3.3.31, both ED110/11. The letter was seven and half pages long.
21. Darré to Alma, 1.3.30., ED 110/10.
22. Darré to Alma, 6.9.32., ED 110/11.
23. Interview with Frau Backe, 14.4.84 and Darré to Alma, 26.3.32, ED 110/11. Lehmann to Darré on Alma, and his feelings about the impending divorce, 21.2.31, NDG/87a.
24. Darré to Alma, 26.6.30, ED 110/10.
25. Darré's diary, 31.1.31.
26. Lane and Rupp, *Nazi Ideology*, pp. 99–103.
27. Otto Wagener, ed. H. Turner, *Hitler aus nächster Nähe* (Frankfurt a. Main, 1978), pp. 250, 252. Darré to Konopath, 'Once you win [Strasser] over, he's a powerful ally', n.d. but August 1930, NDG/87b.
28. Wagener, *Hitler*, p. 212. Wagener claims to have selected Darré for the economic department on Himmler's recommendation, and adds that Himmler told him he and Hess had studied with Darré. This does not square with the version of Himmler's education given by his biographers (he studied in Munich), nor with the picture of the recruitment of Darré that emerges from the correspondence in his Nachlass.
29. Professor H. Haushofer, *Mein Leben als Agrarier* (Munich, 1982), p. 53. There is an interesting account of his first contact with Darré and work with him, in pages 53–61 and 74–77.
30. Ernst Hanfstaengl, mentioned by Lehmann to Darré, 8.12.31, NDG/87a. See Hanfstaengl, *Hitler: the missing Years* (London, 1957), p. 183. His highly coloured memoirs, written for the American market, claim Rosenberg was at least 'half-Jewish', and Strasser and Streicher both 'looked Jewish to me': his argument is that only Jews could propound anti-semitism so vehemently and successfully, pp. 80–1. Class and von Herzberg, see Lehmann to Darré, 5.11.30, NDG/87a.
31. L. and J. Stone *An open Elite?* (London, 1984), and for the volatile land market in East Prussia, see I. Morrow, *The Peace Settlement in the German–Polish Borderlands* (London, 1936), p. 260.

32. W. Abel, *Agrarpolitik* (Göttingen, 1967), pp. 250, 253, and F,. Hennig, *Landwirtschaft und ländliche Gesellschaft in Deutschland, 1750–1956* (Paderborn, 1978), vol 2, p. 149. For a more detailed analysis, see my thesis, Bramwell, 'National Socialist Agrarian Theory and Practice', chapters I and IV.

33. Farquharson, *Plough and Swastika*, chapters I and III, and H. Fallada, *Bauern, Bonzen und Bomben* (Berlin 1931), for a good contemporary account.

34. W. Clauss, 'Erfahrungen aus 50 Jahres Agrarpolitik', unpub. paper given to Schleswig-Holstein *Bauernverband*, 14.7.79, p. 4, and D. Gessner, *Agrarverbände in der Weimarer Republik* (Düsseldorf, 1976), pp. 23 ff, personal interest rates were 9–12%.

35. W. Dawson, *The Evolution of Modern Germany* (London, 1919), p. 256, quotes Freiherr von Wangenheim, the head of the Agrarian League, as calling for state intervention to prevent the private sale and division of estates.

36. Salomon, *Fragebogen*, p. 255.

37. See R. Hamilton, *Who Voted for Hitler* (Princeton, 1982), pp. 38–9, 40–1.

38. See NDG 140, 141, 142, 143, for the memoranda.

39. Quoted in Hamilton, *Who Voted for Hitler*, p. 7.

40. Hans Wehdt, *Hitler Regiert* (Berlin, 1935), pp. 86–7.

41. 'Black blood', Darré's diary, 15.12.33. Marie Adelheid to Charly Darré on Goebbels' pamphlet 'attacking the *Rassenarbeit* of our friends,' 6.10.31, NDG/87a.

42. Darré to Konopath, re Strasser, Goebbels and Himmler, undated letter of August 1930, NDG/87b.

43. Farquharson, *Plough and Swastika*, p. 18.

44. Report of AD activity in Saxony, 1.1.31 to 1.1.32, NDG/140.

45. Hans-Jürgen Riecke, 'Memoirs', unpub. MS deposited in the Coblenz Federal Archives, p. 47.

46. AD Report on Saxony, NDG/140.

47. Darré's diary, 18.12.31.

48. Hamilton, *Who Voted for Hitler*, p. 27, and see E. Fröhlich and M. Broszat, 'Politische und soziale Macht auf dem Lande, *VjhfZg* (1977).

49. Darré's diary, 6.9.32.

50. AD Report to Darré, 28.2.33, NDG/140. The Raiffeisen Co-operatives remain a neglected subject in English. Started in the mid nineteenth century, they flourished largely in Catholic areas, and were independent of State funding, unlike the Schultz-Delitsch co-operatives.

51. *Statistische Jahrbuch* (1934), p. 63.
52. E.g., Eduard David, *Sozialismus und Landwirtschaft* (Leipzig, 1903), and see his running battle with Kautsky's more typically Marxist opposition to small farmers, 'Kritische Bemerkungen zu Kautskys Agrarfrage', *Neue Zeit*, 1890–1900. David Abraham's two chapters on agriculture in the Weimar republic, in the *Collapse of the Weimar Republic* present an interesting but controversial Marxist interpretation of the issue, marred by turgid jargon, and the use of *Landwirtschaft* and *Fraktion* as if they were English words. David Mitrany, *Marx Aginst the Peasants* (North Carolina, 1951), is a splendid and irreplaceable study of Marxism and the agrarian question in Europe. K. Tribe and A. Hussein, however, in *Marxism and the Agrarian Question* (London, 1981), vol I, claim, in a lucid but Jesuitical argument, that David is more representative of the Marxist tradition.
53. See Bramwell, 'Small Farm Productivity under the Nazis,' *Oxford Agrarian Studies*, XIII (1984).
54. E.g., Rundschreiben no. 59, 10.7.31, NDG/142.
55. *Neuadel*, pp. 24–5, and see reference to Habicht, Darré to Alma from Saaleck, 28.34.30, ED 110/10.
56. See page 232, n. 35.
57. Abraham, *Collapse of the Weimar Republic*, pp. 214–5.
58. Rundschreiben no. 59, 10.7.31, NDG/142. One memo. c. 1938, from a farmer to the Reichschancellery, opposing RNS legislation, presented an almost Hayekian argument in favour of 'the will and strength of private initiative, and against bureaucracies, committees and directives', in BA NS 10/104.
59. Sondern Rundschreiben, 9.11.31, NDG/142.
60. Behrens to von Schleicher, 27.1.33, BA R43II/210; von Fleming, President of the National Agricultural Chamber, to von Schleicher, 16.1.33 and 17.1.33, and Darré to von Schleicher, 13.1.33, both BA R43II/192. Von Schleicher placed a quarter of a million marks at their disposal, 25.1.33. See the two telegrams, East Prussian head of organisation of German handworkers to von Schleicher, 13.1.33, BA R43II/192. The minutes of a meeting between the National Agricultural Chamber, the Reichslandbund and the Ministries of Economy and Agriculture discussed the need for emergency measures to combat possible famine among farmers and a fear of 'Communist terrorism.' 11.1.33, BA R43II/192.
61. Bramwell, 'National Socialist Agrarian Theory and Practice', p. 171.
62. The SS reference is in Sondern Rundschreiben, 19.1.33, NDG/140.

See on the early SS, the excellent new study by R. L. Koehl, *The Black Corps* (Wisconsin, 1983).

63. Darré's diary, 20.2.32. For the text of the SS marriage laws of 1931, see Appendix. See also Darré's affidavit on the SS for the *TWCN*, Case 8, and his note on the affidavit, written December 1950 at Bad Harzburg, copies in author's possession.

64. See pages 134, 241, n. 23. For Nazi-Indian links, see the comprehensive study by Milan Hauner, *India in Axis Strategy* (London, 1982).

65. Darré and Himmler made a pilgrimage together to Saxon graves in 1934: surprisingly, Röhm accompanied them, Cecil, *Myth of the Master Race*, p. 96.

NOTES TO CHAPTER FIVE

1. D. Schoenbaum, *Hitler's Social Revolution* (New York, 1966), p. 168.

2. A. Schweitzer, 'Depression and War', *Political Science Quarterly* (Sept., 1947), p. 332.

3. Goebbels, *Diaries*, (ed. Louis Lochner, New York, 1948), p. 165; Grundmann, *Agrarpolitik*, p. 156; Farquharson, *Plough and Swastika*, p. 229.

4. A. Dallin, *German Rule in Russia* 1941–5 (London, 1957), pp. 39–40, 304–319.

5. R. Koehl, *RKFDV: German Resettlement and Population Policy 1939–45* (Cambridge, Mass., 1957), p. 29; A. Speer, *Inside the Third Reich* (New York, 1970), p. 627.

6. H. Backe to Ursula Backe, his wife, 20.8.36, BA NL75 Micr.

7. Backe, 'Disposition', notes written in prison in 1946, copy in author's possession by courtesy of Frau Backe.

8. *Grosser Bericht* (Backe's memoirs, written in prison in 1946, and hereafter referred to as *G.B.*), BA NL75/10, p. 2.

9. *G.B.*, p. 1.

10. Dr Ludolf Haase, 'Aufstand in Niedersachsen' (mimeographed, 1942), chapter 86.

11. J. Noakes, *The Nazi Party in Lower Saxony, 1919–1933* (Oxford, 1971), pp. 22–5.

12. Haase, 'Aufstand', p. 651.

13. 'Wirksamkeit der Skalden', c. 1936, BA NL94II/49, and cf. Pulzer, *Rise of Political Anti-Semitism*, p. 322.

14. Interview with Frau Backe, January, 1980.

15. Noakes, *The Nazi Party in Lower Saxony*, pp. 22, 36.

16. Haase, 'Aufstand', p. 659. Backe was arrested in June 1923 for bill-posting, and delivered caustic speeches at the police station, comparing his treatment with that in Russia. But the timing of the arrest was fortunate. The inflation enabled him to pay a heavy fine of 16,500 marks within weeks, BA NL75 Micr. police papers and copy of bill, Haase, 'Aufstand', pp. 257, 257a. The view on Russia held by Backe was common in the circle round *Die Tat*, the neo-Conservative, *völkisch* journal to which Backe subscribed. Backe was a friend of Ferdinand Fried Zimmerman, economist and journalist, and later a Nazi publicist. Dr Giselher Wirsing, another *Die Tat* writer, and later Nazi intellectual, stressed the concept of *Zwischeneuropa*—Germany's influence should extend into Russia. See Klemens von Klemperer, *Germany's New Conservatism* (Princeton, 1957), pp. 130-4, 205; Zimmerman to Backe, 30.4.36, BA NL75/10; Report by Dr Wirsing, end 1942, 'Westlicher Imperialismus oder deutsche Ordnung im Osten', BA R7/2243.
17. Speech to the Worker and Middleclass Union, *Göttinger Tageblatt*, 17.1.23.
18. G. Stoakes, 'The Evolution of Hitler's Ideas on Foreign Policy, 1919-1925', *The Shaping of the Nazi State*, ed. P. Stachura (London, 1978).
19. Backe's speech, 'The Foreign Policy Aims of the Young German Order', early 1925, BA NL75/6. Backe's rejected thesis was to be published in 1942. Noakes, 'Conflict and Development', p. 25, suggests that 1925 plans for eastern expansion and development submitted to a conference of the Göttingen branch of the NSDAP were by Backe, and 'read like a blueprint for later SS policy'; however, Backe appears to have cut his connection with the NSDAP by this time, and did not attend the conference. See also *G.B.*, p. 40.
20. Speech, and notes for speech, c. 1925, BA NL75/5. Haase and Fobke criticised Strasser's 1925 programme for its urban and parliamentary emphasis, BA NS26/896.
21. Haase, 'Aufstand', p. 644, and cf. Lane and Rupp, *Nazi Ideology*, pp. xiii, xviii.
22. Ibid.
23. Halford Mackinder, the English geographer, appears to have inspired German geo-politicians, and in *Democratic Ideals and Reality*, first published in 1919, the reprinted version (New York, 1942), referred to the use made of it by General Karl Haushofer.
24. Darré to Backe, 12.8.31, Backe to Darré, 23.8.32, Darré to Backe,

12.9.31, and see *G.B.*, p. 11. letters in Frau Backe's possession. See 'Volk and Landwirtschaft', 1931, and 'Deutsche Bauer Erwache', 1934, for Backe's response to Darré's influence.

25. Darré's post-war account, BA NL94I/28.
26. *G.B.*, pp. 4, 11, pp. 20–30. Hitler's speech printed in M. Domarus' *Hitler; Reden und Proklamation 1932–45* (Munich, 1965), p. 60.
27. Backe to Frau Backe, 5.4.33, BA NL75 Micr.
28. Postcard, 2.1.33, BA NL75 Micr. BA NL75/6, *G.B.*, p. 12.
29. Darré's diary, 20.4.33; J. Heinemann, 'The Resignation of Hugenberg', *JMH*, 41 (1969), describes Hitler's manoeuvre, and see also J. Leopold's interesting biography of Hugenberg, *Alfred Hugenberg; the Radical Nationalist Campaign against the Weimar Republic* (Yale, 1978).
30. See Grundmann, *Agrarpolitik*, pp. 60–2; Farquharson, *Plough and Swastika*, p. 68, 52, *G.B.*, p. 14, A. Schweitzer, *Big Business in the Third Reich* (Bloomington, 1964), p. 197. Von Rohr's criticisms are in 8.4.33, NDG/143.
31. 8.4.33, BA R43II/192, and Backe to Frau Backe, 21.3.33, BA NL75 Micr.
32. Darré's telegram to Hitler, 16.4.33, BA R43II/192, and H. Kehrl, *Krisenmanager im Dritten Reich* (Düsseldorf, 1971), p. 49, on Darré's suitability for the post.
33. 'Hugenberg belongs to the year 1900', commented Darré sourly in his diary, 20.4.33.
34. Grundmann, *Agrarpolitik*, pp. 34, 36.
35. Ibid.
36. Ibid., note on p. 169.
37. Darré's diary, 10.4.33; Backe to Frau Backe, 3.5.33, BA NL75 Micr.,
38. Darré's diary, 18.5.33, and interview with Frau Backe, 14.4.84.
39. See pages 107–8.
40. Backe to Frau Backe, 6.9.33, BA NL75 Micr., and 24.10.34, Frau Backe's diary.
41. The photo album with the pictures of Goslar is in Frau Backe's possession. The personal files are kept at the Federal Archives, Coblenz.
42. Backe to Frau Backe, 6.9.33, BA NL75 Micr.
43. The annotated copy of the *RGB* is in Backe's microfilmed papers. See also Frau Backe's diary, 14.10.33, and Darré's diary, 29.9.33. Gürtner to Lammers, BA R43II/193, discusses the legislation.
44. Schoenbaum, *Hitler's Social Revolution*, chs. 1, 8, 9.

45. R. W. Darré, 'Ostelbia', repr. in *Blut und Boden* (Munich, 1941), 20.5.33, BA R43II/192, Haushofer, *Ideengeschichte*, p. 275, and Farquharson, *Plough and Swastika*, p. 203.
46. Backe to Frau Backe, 14.10.33, BA NL75 Micr.
47. See Schweitzer, *Big Business*, pp. 165, 167.
48. Meeting with Hitler, *G.B.* p. 38, Frau Backe's diary, 17.7.34.
49. See report on food supply and agricultural prices, 18.9.36, BA NS10/54, and report on agricultural income, Institute of Economic Research report, 1936, BA R43II/194. Dr T. W. Mason argues that official figures do not reflect the true rise in retail food prices, 'National Socialist Policies towards the Working Classes', Oxford D. Phil Thesis, 1971, pp. 169–172. Statistics are in *St. Jb. f.d. deutsches Reich*, cited in Tracy, *Agriculture in Western Europe*, p. 207, and see Guillebaud, *The Economic Recovery of Germany*, p. 155, Frick's 21 reports, 24.7.35 and 2.8.34, on fats prices, in BA R43II/193.
50. Frau Backe's diary, 17.7.34.
51. Backe's meeting with Ley, Dr Krohn, and the Trustees of Labour, 27.8.34, see BA R43II/318.
52. Goering's attack, 31.8.34, BA R43II/193; Goebbels on *Winterhilfe*, 15.12.33, Darré's rebuttal, 17.1.34 and 28.3.34, BA R43II/1143.
53. Farquharson, *Plough and Swastika*, p. 100.
54. Backe to Frau Backe, 5.7.35, BA NL75 Micr.
55. Ibid., 2.8.36, BA NL75 Micr., G.B., pp. 13–20.
56. See Hitler Decree, 4.4.36, BA R26 I/35, G. Thomas, *Geschichte der deutschen Wehr-und Rüstungswirtschaft, 1918–1943–5* (Coblenz, 1966), pp. 111, 124, ·15.5.36, BA R26I/36, and minutes of meeting on raw materials and currency, nineteen ministers present, 4.5.36, BA NL75/9; Backe to Lammers concerning Backe's membership of the committee, 15.6.36, BA R43II/208, Parchmann to Backe, 10.5.38, BA NL75/10, and notice of 6.7.36, BA R26/I/1a. Darré's comment on Körner, 25.8.41, BA NL94II/20.
57. Schweitzer, *Big Business*, p. 200.
58. Backe's notes on his meeting with Goering, summer 1936, BA NL75/5, and Frau Backe's diary, 25.11.36.
59. Backe to Frau Backe, 25.11.36, BA NL75 Micr. Darré's diary does not record a talk with Himmler at this time.
60. See unintentionally hilarious account in BA R16/2141, correspondence August 1935. The KBF refused to stop repeating the allegations, despite repeated roughing up by party officials. Darré's warning about *Die Tat's* Masonic connections, Rundschreiben 110, 19.12.31, NDG/142.

61. 1.11.36, NDG/483. Goering had circulated the national leadership for their views on Germany's economic situation.
62. For example, Darré always insisted on being addressed in writing as R. Walther Darré, never Richard, much less Ricardo Darré. According to him, this was to avoid confusion with his father, while his father lived. According to his colleagues, it was to emphasise the name 'Walther' (etymologically close to the German word 'to rule') rather than 'Ricardo', with its Latinate origins. He became known to many of his peers as 'R punkt Walther'.
63. Hitler's speech, quoted in *Landpost*, 18.9.36, in NDG/315. See Backe to Frau Backe, 10.9.36, BA NL75 Micr, and Darré's diary, 4.2.38; Körner to Backe, 7.9.36, BA NL75/8.
64. *G.B.*, pp. 17, 21, 46.
65. See *Molkerei-Zeitung*, 20.8.36. Backe to Frau Backe, 12.5.37, discusses Goering, BA NL75 Micr.
66. Backe to Frau Backe, 20.8.36, BA NL75 Micr.
67. Darré on meeting of 3.2.37, 16.2.37, NDG/137. See also Darré's diary, 5.2.37.
68. Darré's memos to Lammers on these subjects are surprisingly clear and well argued. See, e.g., BA R43II/611.
69. *G.B.*, p. 20.
70. Interview with Frau Backe, 14.4.84.
71. Backe's complaint about Darré in 1940, see Frau Backe's diary, 19.6.40. See also Haushofer, *Ideengeschichte*, pp. 270–1. See chapter Eight for a discussion of Darré's campaign for organic farming.
72. See chapter Six, Darré's diary, 30.7.37, and his memorandum to Goering, protesting about the effects of the Pact, 29.9.39, BA NL94II/20.
73. 1.4.38 and 16.3.39, Darré's diary, 1.4.38 and 16.3.39.
74. For the peasant settlement issue, see Bramwell, 'National Socialist Agrarian Theory and Practice,' Chapter V.
75. See Darré's memorandum, calling for fewer imports of Polish labourers, 16.11.39, BA R2/31089, and his diary, 30.7.37. Dr Mason's comment is in his *Sozialpolitik im Dritten Reich* (Opladen, 1977), pp. 228. See also pp. 158–9.
76. Kehrl, *Krisenmanager*, p. 82. Hitler's complaint was reported by Eggeling, then Keeper of the Seal at Goslar, in a file note 7.4.37, BA R16/2222.
77. Darré's diary, 3.5.38.
78. See FKG, *Reichsgesetzblatt* 3 (1938), pp. 825–32, *G.B.*, p. 35, but cf. D. Schoenbaum, 'Class and Status in the Third Reich', Oxford D.

Phil. Thesis (1964), who argues that the estates were mostly forest anyway, and estate ownership hardly affected by the FKG.

79. Darré's diary, 4.3.39, and *G.B.*, pp. 35–45. For the draft new Ministry, see Darré's diary, 31.1.39.

80. See chapter Six, and, for Backe's criticisms of Germany's sloth, his draft 'Testament', 1946.

81. See pages 96–7.

82. 1943, see Backe's draft for a speech in Pomerania, 12.7.43, BA NL75/9, and his jottings, c. 1943, BA NL75/10. For the refusal of permission for his wife to travel, see Backe to Frau Backe, 4.10.44, BA NL75 Micr.

83. 10.1.41, Frau Backe's diary. See also *G.B.*, pp. 39 and 46, and Grundmann, *Agrarpolitik*, pp. 73–4. Secrecy from Darré, see Farquharson, *Plough and Swastika*, p. 247, and 'Row with Darré over secrecy of Russian preparations', Frau Backe's diary, 27.6.41.

84. Darré warned Backe against Bormann, *G.B.*, pp. 39–40; for Darré's attitude to Bormann, see J. von Lang, *The Secretary; Martin Bormann, the Man who Manipulated Hitler* (New York, 1979), pp. 51–2.

85. Erich Dwinger, *Die 12 Gespräche, 1933–1945* (Germany, 1966), pp. 50–54.

86. Darré's letter to Backe, 14.3.41, BA NL75/10, Frau Backe's diary, 5.12.39 and 19.6.40. Hitler's request is mentioned in Frau Backe's diary, 28.10.40.

87. Backe to Frau Backe, 8.4.41, BA NL75 Micr.

88. See G. Denschner, *Heydrich, the Pursuit of Total Power* (London, 1981), p. 286, who also makes the point about Heydrich and Backe. For Hayler, see Backe to Frau Backe, 15.11.43, BA NL75 Micr, and Hayler to Frau Backe, 10.4.44 and 30.4.44, BA NL75/10. See also interview with Frau Backe, July 1981.

89. Moritz' file note of Darré's phone call, 30.8.41, BA R14/371, Darré's accusation that Backe was falsifying the figures, 29.8.39, Darré's diary, and Albrecht to Brückner (a Hitler adjutant) on Darré's attack on the figures, 1.6.40, BA NS10/107. Backe's report on European food self-sufficiency is in BA NS10/107.

90. Frau Backe's diary, 5.11.41. Cf. Brandt, *Fortress Europe*, p. 131, on the Germans' failure to find vast grain reserves in Russia. For the withdrawal in 1943, see Brandt, op. cit., pp. 148–9, and Backe to Lammers, 9.4.44, BA R43II/614. For the frozen potato harvest, see report from Danzig to Reich Chancellery, 16.12.41, BA R43II/863.

NOTES TO CHAPTER SIX

1. R. Cecil, *Rosenberg, The Myth of a Master Race* (London, 1968), pp. 119–20 and *passim* for an interesting account of the links between Rosenberg, Darré and Himmler. Cecil considers that Darré was a potent influence on Himmler in the early '30s, awakening long-dead enthusiasm for the soil and for settlement. I. Ackermann, in *Heinrich Himmler als Ideologe* (Göttingen, 1970), thinks that Himmler's early interest in eastern settlement affected Darré. However, this may have been over-estimated as a factor in Himmler's early ideology. Darré wrote his books before meeting Himmler.

2. 19.1.38, Darré's diary.

3. I. Kershaw, 'The Führer Image and Political Image', in ed. Hirschfeld, G., and Kettenacker, L., *Der Führerstaat; Mythos und Realität* (Stuttgart, 1980), estimates that some 80 per cent of Germans continued to support Hitler, if not the Nazi party, up to 1944. Although *Mein Kampf* was a best seller, not many buyers read it all the way through.

4. See BA NS2/108 and NS2/85.

5. The estimated 11.5 million ethnic Germans living in Europe outside the boundaries of the *Altreich*, Sering, *Agrarverfassung*, p. lv.

6. *St. Jb.* (1941), and *Landesbauernschaft in Zahlen* (1938–44).

7. K. Meyer, 'Lebensbericht' (unpub. MS., 1970, in author's possession), p. 95.

8. Kummer's doctoral dissertation of 1919 concerned Eastern settlement. The Archive for Internal Colonisation was founded in 1908 by Prof. H. Sohnrey, who later edited *Neues Bauerntum*. Quote from Meyer, 'Lebensbericht', pp. 108–9.

9. K. R. Schultz-Klinken, 'Preussische und deutsche Ost-siedlungspolitik von 1886–1945, Ihre Zielvorstellungen, Entwicklungsplan und Ergebnisse', *ZAGAS*, 21 (1973), 198–215.

10. 'Die Ansiedlung deutscher Bauern in den eingegliederten Ostgebieten', *Vortrag* by Prof. Dr Otto Auhagen, undated, BA R49/20, and see p. 139, this chapter. Fischer, F., *Griff nach der Weltmacht* (Düsseldorf, 1967), pp. 104–5, 138–9, 142–3, 234–5, cited by Farquharson, *Plough and Swastika*, p. 257.

11. See Broszat, Martin, *NS Polenpolitik 1939–45* (Stuttgart, 1961), pp. 13–16, and Rich, N., *Hitler's War Aims*, ii, (London, 1974), pp. 70–71.

12. E.g. see correspondence between Pancke, head of RuSHA after Darré's resignation, and General von dem Bach in Breslau in November 1939, which discussed at length the bad relations

between the SS and the Gauleiters of Silesia, West Prussia and the Warthegau, who had complained about the SS to Hitler, 27.11.39, and later correspondence, BA NS2/60. See also attacks on SS plans to form land units by the SA in 1937, BA NS2/290.

13. L. Wheeler, 'The SS and the Administration of Nazi Occupied Europe 1938–1945' (Oxford D. Phil. thesis, 1981), 'Although Darré was removed from his post by Himmler in 1938, the RuSHA which had developed under his auspices was a powerful component of the Reichführer's network', p. 27. This *became* true in 1940–41, but by 1943 the RuSHA had lost its importance: it was certainly not all powerful in 1938.

14. R. Koehl, *RKFDV; German Resettlement Policy, 1939–45* (Harvard, 1957).

15. RK26272 B, Führer Decree, 7.10.39, BA R49/1 and 4.

16. See Wheeler, 'The SS', pp. 12–17, 20, on the administration of the RKFDV.

17. Backe, *G.B.*, p. 46.

18. See Loewenberg, P., 'The Unsuccessful Adolescence of Heinrich Himmler', *AHR* (1971), and the subsequent criticism by Novak, Steven J., *AHR* (1972). Despite 'the careful scholarship of Smith and Angress and Ackerman ... it has not been possible so far to identify the political ideas of Heinrich Himmler before 1933'. Lane and Rupp, *Nazi Ideology*, p. 156.

19. Himmler to Darré, enclosing a draft order referring to the first SS settler selection camps set up in December 1935, 30.4.36, BA NS2/50. See also Darré to Himmler on 30.4.36, BA NL94 II/49.

20. 18.12.37, Darré's diary.

21. See pp. 141–145, this chapter. Backe warned Darré in 1938 that Himmler was anxious to take over settlement activity, *G.B.*, p. 40.

22. Report of speech by Himmler, marked 'Secret; E. P. Madrid', 20.10.40, BA R49/20.

23. 'I intend to fight and settle ... some day, far from lovely Germany' (Himmler's Diary, 11.11.19). Not much is known of Himmler's life between 1924–30. His 1922 plan to emigrate to Turkey never re-emerged, but he befriended Indian nationalists who were studying in Germany during the 1920s. See BA NS19 neu/622, 1068, 1812. See correspondence 1924–30 with Taraknath Das, Indian revolutionary, in Georgetown University, Washington. For the emigration to Turkey plan, see the comment in Angress and Smith, 'Diaries of Heinrich Himmler's early years', *JMH*, 31 (1959), 206–225.

24. E.g. Ackermann, *Himmler als Ideologe*.
25. See Meyer, 'Lebensbericht', p. 108.
26. Eggeling to Himmler, 23.10.40, BA NL75/10. Eggeling sent a copy to Backe. Darré had written in 1939, 'Now I cannot look after the peasantry's interests any more. Among the rural population this will presumably make me neither renowned nor loved. I can only hope that History will justify me to the grandchildren of today's peasants', 13.5.39, Darré's diary.
27. See Haushofer, *Ideengeschichte*, p. 313, 2.2.40, Darré's diary, and whole file, BA NL94 I/1d.
28. E.g. 7.9.40, BA NS10/37, and BA R43II/646b, pp. 52–3.
29. Given in full in the Appendix.
30. RFSS Order on the formation and membership of RuSHA, in BA NS2/99 and NS2/1, pp. 3–4. Details of the education courses for young SS leaders are in BA NS2/108. Himmler requested that educational programmes should be 'soldierly, not dry, academic stuff ... geography, history or racial education, recognition of enemies (*Gegner*), Freemasons, Catholics, etc.', 4.10.37, BA NS2/85. Oddly enough, the list of opponents did not include Jews. Darré's younger brother Erich attended the courses, and by 1937 was SS SBF in the RuSHA library department, 23.3.37, BA NS2/100.
31. BA NS2/99.
32. BA NS2/100.
33. Bulletin for 1934–5, BA NS2/1.
34. Order dated 15.10.34, BA NS2/99. Himmler took the rune from *Odal*, the title of Darré's journal.
35. The introduction to the RuSHA file at the Federal Archives, Coblenz gives the office's history.
36. Himmler to Dr Reischle, 12.3.39, BA NS2/66.
37. M. Kater, 'Die Ahnenerbe', *VjhfZg* (1974).
38. In Case 8 of the *TWCN* the RuSHA was charged with expropriating all enemy and Jewish property in areas occupied by German troops. However, the papers that have survived seem to concern straightforward purchase and settlement procedures in the incorporated areas, the Danubian plain, the Alps, Lorraine and other 'Germanised' areas. The usual instrument was the *DAG*, Germany's largest settlement society, taken over by the SS in 1938; see p. 141 above.
39. See e.g. Darré's circular to the NSDAP AD in 19.1.31, where he informed his members that the SS was asking for volunteers,

34. Priebe, Hermann, 'Wirtschaftsziele eines Umsiedlerhofes im Warthegau'. *Sachfragen und Tatbestände*, in *Neues Bauerntum* (Jan. 1941), p. 17.
35. 12.2.41, BA R49/1/39. Typically, the report also attacked the old Prussian Agricultural Commission: 'It would have needed forty years to do what we have done in one'.
36. June 1940, BA R49/20.
37. 16.6.40, BA R43II/223a.
38. The Germans from the South Tyrol, who were settled in Southern Galicia in 1939–40, also come into this category. There is a good coverage of the ethnic German problem in Sering, *Agrarverfassung*. Between 1910 and 1934, the German population of Posen and Pommerellen fell from 1.1 m to 312,690. In August 1939, 70,000 recent refugees from Poland were in transit camps in Germany. Meyer, 'Lebensbericht', p. 102. See also Schieder, *Documents on the Expulsion*: Proudfoot, *European Refugees 1939–51*, (London, 1957). The economic agreement signed with Russia, 19.8.39, was also important. Rich, *Hitler's War Aims*, i, p. 128.
39. Richard Bessel, 'The SA in the Eastern Regions of Germany, 1925–30' (Oxford, D.Phil. Thesis, 1980), ch. 1.
40. See Judgment in Darré's trial, *TWCN*, xiv, 558–60.
41. See Kulischer, Eugene, M., *The Displacement of Population in Europe* (Montreal, 1943, pub. International Labour Office), Map I. 93,000 Bessarabians; 21,000 Dobruja Germans; 98,000 Bukovinians; 68,000 Wolhyniens; 58,000 Galicians; 130,000 Baltic Germans; 38,500 E. Poland; 72,000 Sudeten Germans; 13,000 Germans from Slovenia: a total of 591,500 people. See also H. Weiss, 'Die Umsiedlung der Deutschen aus Estland', *JBD* (1964).
42. See Koehl, *RKFDV*, pp. 146 & 152, on Himmler and the German plan in the East. Himmler tried to transcend his power limits in 1942, p. 147. Cf. Dallin who in *German Rule in Russia*, p. 255, discusses plans for resettling the Crimea in 1942.
43. Sering, *Agrarverfassung*, pp. 177–208, especially p. 183.
44. See Wiseley, W. C., 'The German Settlement of the Incorporated Territories of the Wartheland and Danzig West-Prussia, 1939–45' (London, Ph.D. Thesis, 1955), pp. 39–41, 43–9, 50–2.
45. See Broszat, *NS Polenpolitik*, p. 47, who puts the figure of Germans murdered just after the German invasion at 6 to 7,000. This controversy has received little publicity outside Germany, as Goebbel's propaganda trick just before the invasion has given the impression that all Polish activity directed against Germans was

Nazi propaganda. Nicholas Bethell, *The War Hitler Won* (London, 1972), completely discounts the idea of large-scale Polish anti-German measures, and accepts the '300' figure, pp. 84, 143. See also Rich, *Hitler's War Aims*, i, 129, and Koehl, *RKFDV*, p. 44. H. H. Krausnick and H. Wilhelm, *Die Truppe des Weltan-schauungskrieges: Einsatzgruppen der SS & des SD 1938–42* (Stuttgart, 1981), p. 55, estimated a minimum of 1,000 German dead in Bromberg and 4 to 6,000 German dead a few days later; Wiseley, 'The German Settlement', p. 108, estimates 10,000, mostly saboteurs.

46. Sering, *Agrarverfassung*, p. 158.
47. 78,000 sq. km of Danzig-West Prussia, of which 47,000 had been German before the First World War; 7,800 sq. km of East Upper Silesia; 17,000 sq. km of the Zichenau was annexed to East Prussia. Brandt, *Fortress Europe*, p. 36. The *St. Handbuch* (1948), gives figures of 90,000 sq. km and 9.9 m. people. Cf. Broszat, *NS Polenpolitik*, pp. 31–6.
48. Erlass des Führers und Reichskanzlers zur Festigung deutschen Volkstums, 7.10.39, Rk 26272 B, BA R49/12. Cf. Wheeler, 'The SS', Ch. 1 & 2; Koehl, *RKFDV*, pp. 31–2.
49. SS Sturmbannführer Brehm to the leader of the RuS Central Settlement Office, from Rielitz, 21.10.39, BA NS2/60.
50. Pancke to Himmler, Bericht der Reise zu dem Höheren SS und Polizeiführer in Danzig und zu dem Höheren SS und P-F in Posen', 20.11.39, BA NS2/60, and see Koehl, *RKFDV*, pp. 42–3.
51. 'Bericht', Pancke to Himmler, 20.11.39, BA NS2/60.
52. Note of a meeting, 29.11.39, between Pancke, Himmler and v. Holzschuher, 1.12.39, BA NS2/60. By December 1939, Greifelt was signing directives for Himmler, Koehl, *RKFDV*, p. 64.
53. Pancke to Himmler, 23.12.39, talk with Himmler, 27.12.39, BA NS2/60.
54. Pancke to Himmler, enclosing a report by the RuS Advisory Staff, 20.11.39, BA NS2/60.
55. Himmler to Pancke, 28.11.39, BA NS2/60.
56. Himmler to Pancke, 12.12.39, and Pancke to Greifelt, 27.12.39, BA NS2/60. See Krausnick, and Wilhelm, *Die Truppe*, on the role of the *Einsatzgruppen*.
57. Pancke to Hildebrandt, 27.11.39, BA NS2/60. An undated report from the RuS to Himmler claimed that 'many urgently necessary repairs had not been carried out because the competence of the *Reichskommissariat* was being attacked. This fight over whose job it is reaches right down to the lowest office positions'. BA NS2/56.

58. Von dem Bach to Pancke, 27.11.39, BA NS2/60. See Koehl, *RKFDV*, p. 245, for Pancke's later career in Denmark as Higher Police Chief and SS Führer.
59. File note of a report by SS GF Hildebrandt, 26.11.39, BA R75/13. Figures on expected returnees are from a report by SS SF Hoffmeyer, 23.5.40, BA R49/20.
60. This was in response to criticism by the RNS in June 1940, when *NS Landpost* ran a series of highly critical articles on the re-settlement; Pancke to Himmler, 3.7.40, BA NS2/56. It was probably not just an excuse. Himmler talked of 'eventual' re-settlement of 150,000 Bessarabian and Lithuanian Germans in Feb. 1940, 6 months after the Ribbentrop–Molotov Pact, Koehl: *RKFDV*, p. 82.
61. SS SF Hoffmeyer, Report, 23.5.40, BA R49/20. Hoffmeyer also noted that conflicts of competence made negotiations difficult in Russian occupied Poland. 'If something is ordered by the NKVD, the military will overrule it within an hour, and the local Soviet will then alter it again ... In general, the NKVD is the strongest authority', p. 2.
62. See Löber, Dietrich, A., ed. *Diktierte Option: Die Umsiedlung der Deutsch-Balten aus Estland & Lettland 1939–41* (Neumünster, 1972), which stresses the improvisatory nature of the re-settlement.
63. Pancke to Himmler, undated report, pp. 127–32, BA NS2/56. For orders regarding the *VD* see *Anordnung*, 24.11.39, 9.5.40, 10.8.41 and 15.12.41, where Himmler orders re-settlement, R49/4. See also 'Die Organisation der Lager und die Aufnahme der *VD* in den Lagern erfolgt durch die *Vo-Mi-Stelle*', BA R49/20. 'Pocket money for Estonian, Latvian, Wolhynien, Galician, Narewian, S. Tyrolean only. To be paid to those in assembly camps, and those who cannot earn for themselves ... *Übergangsgelder* to be paid to Baltic Germans not in camps. 5 RM for adolescents, 3 RM per child', BA R49/12. See circulars on labour units to be formed from *VD* in *Altreich* camps, in BA R16/17.
64. On evacuation see Broszat, *NS Polenpolitik*, p. 87 and Gross, *Polish Society*, p. 71. Report, 3.7.40, BA NS2/56, esp. pp. 127–32. Evacuation order for Polish dwarf farms 1942; 1 wagon per household: (a) must take: important papers, clothing, food for five days, gold and valuables, washing materials; (b) can take: bed linen, bicycles, 1 animal p.c. (pet), *Einzelmöbel*; (c) NOT to take: flour, fruit, fresh vegetables, agricultural machines, bees, large animals, dogs. BA R75/10.
65. *Sitzung* no 35 re: Bessarabian Germans. Greifelt's speech in Danzig

called for them to be given jobs in industry, 3.12.40, BA R49/12.

66. RFSS, *Schnellbrief* to administrators within the conquered territories, 12.1.40, BA R49/12.

67. Pancke to Himmler, undated report, c. Aug. 1940, BA NS2/56. See also Bannister, Sybil, *I Lived under Hitler* (London, 1956), which describes life in wartime Bromberg and the surrounding countryside, and gives a picture of two mutually hostile groups in the *Volksdeutsche* and *Reichsdeutsche*; even the tennis club membership was split down the middle. In a report to Himmler, Pancke describes one Polish estate:

> The owner, a senator, was shot some time ago for enmity to Germans. The son makes an impression of extremely low racial value, with a mongolian-tartar face ... speaks only French with his mother. The German son of the former owner, Count Schoenberg, we found on the farm, in civilian dress ... eating with Poles at table. He should not take over the farm.

20.11.39, BA NS2/60. See also Wiseley, 'The German Settlement', pp. 256–7, pp. 263–4: Polish workers and peasants hostile to Polish nationalism.

68. Melita Maschmann, *Fazit, Kein Rechtfertigungsversuch* (Stuttgart, 1963), p. 100.

69. Report by *Vo-Mi Stelle*, 23.5.40, BA R49/20. Report 'Ein Jahr Deutscher Ostsiedlung', December 1940, by Bronia Alix Elsas, of the SS Staff Office, on the '*Nahplan 3*', BA R49/1/34.

70. Elsas report, R49/1/34.

71. Koehl, *RKFDV*, p. 73. 12,000 Balts were settled 'with difficulty' in towns. See Greifelt to *Einwanderungstelle* Lodz, asking for information on the size of the former homes of the Baltic Germans, 12.5.40, BA R49/12, and *Landrat* Krotoschin to Governor in Posen, 'There are not enough large farms suitable for Balts ... please do not send any more Balts ... I can only settle peasants here', 1.3.40, BA NS26/943. A report by W. Quering to German Foreign Institute in Stuttgart described Greifelt's speech on the ethnic Germans as '*Kopfwasche*', 12.4.40, BA R49/20.

72. Kruger to Frank, 25.3.41, BA R52/11/238, p. 7. Cited Wheeler, 'The SS', p. 101. See also the annual report of the DAG, 'Because of the shortage of labour, settlement activity remained confined almost entirely to individual farms, and the continuation of

projects undertaken last year encountered great difficulties', 31.12.41, BA R49/118, p. 1, cf. Rich, *Hitler's War Aims*, ii, 83: Himmler ordered evacuations to cease in May 1943.

73. Report 1941, 10.2.41, BA R49/1/34.

74. Broszat, *NS Polenpolitik*, p. 97, 'increasing pressure and resistance among Polish population after evacuations'. See also SS Planning Staff, Posen, Report, pp. 43–47, and p. 73, 12.2.41 and 1.4.41, BA R49/1/34, and reports for 8.5. 41 and 1.9.41 in BA R49/120.

75. 'Lage und Stimmung der Polen des Umsiedlungsgebietes'. Report by SS BF U. Greifelt, EWS Lodz, 26.3.41, BA R75/3. He gave examples of Poles offering hot milk for the children of the German refugees, amidst 'outbreaks of hate', such as remarks like 'You Germans have not won the war yet, we have not lost Poland'. But 'Several sources ... state that the German victory was received with joy by the peasants in certain areas ... in many ethnically Polish villages'. In general, Germans were received with least hostility in the countryside. Gross, *Polish Society*, pp. 128 and 140.

76. Cf. Koehl, *RKFDV*, p. 122 who quotes Frank's diary: 210,000 voluntary labourers by April 1940. Gross, *Polish Society*, p. 78 estimates that some 100,000 Polish workers volunteered in 1939/40, and that the final proportion of volunteer workers was 15 per cent. Agricultural work was the most unpopular.

77. SS Ansiedlung Stab Abt. 1, Report by SS BF Schelpmeyer, 24.1.42, BA R49/1/35. Polish labourers to *Altreich*, Broszat, *NS Polenpolitik*, pp. 102–4.

78. Brandt, *Fortress Europe*, p. 41. Vegetable production increased 30 times, ibid., pp. 42, 49.

79. Report by R. M. Schmidt, *Aussenstelle Ost*, 8.5.41, R49/Anhang 1/ 39.

80. Priebe, H., 'Wirtschaftsziele eines Umsiedlerhofes im Warthegau', *Neues Bauerntum* (Jan. 1941), BA NS26/948.

81. E.g., Himmler wrote of the need to make the Eastern landscapes which currently were 'identical to those of Bronze Age Europe' come to resemble the cultivated fields of Schleswig-Holstein, 21. 12.42, BA R49/4.

82. Himmler, cited in Wheeler, 'The SS', pp. 96–7.

83. Correspondence 1942–3, re *VD* in transit camps in the *Altreich*, and their recruitment for agricultural labour units, BA R16/164. In June 1940, Greifelt called on German businessmen to offer jobs to the Bessarabian Germans, BA R49/12.

84. DAG report of 1950, BA R49/25, p. 51.

85. R. Koehl, *RKFDV*, pp. 222–3. See Broszat, *NS Polenpolitik*, p. 41; 10,000 Poles and Jews had been murdered by Feb. 1940. Documents from the *Einwanderungsstelle*, Lodz contain details of the evacuations, BA R75/10 and R75/13.

86. On the psychological effects of the emptier, wilder countryside, see Maschmann, *Fazit*, p. 104, but also Hohenstein, *Tagebuch*, pp. 299–300, who found the untouched emptiness of the 'new area' stimulating, as he rode around it. Broszat describes the effect on the *Volksdeutsche* of the sight of the dispossessed Poles, *NS Polenpolitik*, p. 97. See also Wiseley, 'The German Settlement', p. 218.

87. See (unsigned) report on Bessarabian resettlement, pp. 14–20 (in author's possession). Sering, *Agrarverfassung*, p. 202. 'The Wolhynien German farming methods are still somewhat primitive ... there was a strong emphasis on grain production'. See also Koehl, *RKFDV*, p. 99, who argues that they were a liability as trustees of efficient, progressive farms in the Warthegau.

88. Löber, Dietrich, *Diktierte Option*, p. 60, calculated that 20 per cent of the Estonian and Latvian German settlers are known to have been killed; this figure did not include missing. See ibid., Part 3, pp. 633–97 for some post-war documents on this subject. The Volga Germans in Russia probably suffered a worse fate. See also Schieder, *Documents on the Expulsion*, G. Trittel, *Die Bodenreform in der Britischen Zone, 1945–49* (Stuttgart, 1975), p. 20. 8 million refugees had arrived from the East by Spring 1947. De Zayas estimates a total of 15 million German refugees. *Nemesis at Potsdam* (London, 1979), p. xix.

89. E.g., 'When one considers, besides, that the farms were ripped from their former owners, then morally speaking one can only welcome the quick end of the episode through the end of the war', Dr W. Lenz, 'Erbhöfe für baltische Restgutsbesitzer im Warthegau', *JBD*, xxix (1982), 127.

90. Gross, *Polish Society*, pp. 35–41.

91. See e.g. Krausnick & Wilhelm, *Die Truppe*, and Christian Streit, *Keine Kameraden* (Stuttgart, 1978), for SS responsibility and the *Einsatzgruppen* during the war.

92. Gross, *Polish Society*, p. 197.

93. Koehl, *RKFDV*, pp. 80ff. 'The ascriptive interpretation of race had to be abandoned little by little', Gross, *Polish Society*, p. 196. See also Struve, *Elites Against Democracy*, pp. 423–5, on the expedient and 'open-ended' character of Himmler's racial criteria, Koehl, *The Black Corps*, p. 215.

94. See correspondence from Brandt, Himmler's adjutant, to Backhaus, Backe's personal assistant, 15.11.44 to 26.2.45, BA NS19neu/1004.
95. See Wheeler, 'The SS', Brandt, *Fortress Europe*, pp. 74, 78, chart of the various authorities in the East, H. Buchheim, 'Rechtstellung und Organisation des RKfdFdV', *Gutachten* des IfZG (Munich, 1958), Eisenblätter, S., 'Grundlinien der Politik des Reiches gegenüber dem G–G 1939–1945' (Frankfurt/M, Diss., 1969), Graf, H., 'Zum faschistischen Okkupationspolitik des deutschen Imperialismus in Polen während des 2. Weltkrieges' (Berlin, Diss., 1961).
96. Rich, *Hitler's War Aims*, i, disagrees, 'When the war was at its height in Russia ... peasants were shifted *en masse* from one end of Europe to the other ...', pp. 57–8.
97. See Koehl, *RKFDV*, 'instead of real overall policy, the *ad hoc* solutions to immediate problems became the major task of the *RKFDV*,' p. 225.
98. Brandt, Himmler adjutant, to Pancke, RuSHA, 24.2.40, BA NS2/55.
99. Koehl, *RKFDV*, pp. 83–4, cannot find evidence of settlement planning East of the incorporated areas after November 1941, 'Meyer's far-flung projects, resubmitted to Himmler in December 1942, were cut off in embryo by the fortunes of war', p. 159.
100. Dallin, *German Rule in Russia*, pp. 285–8, esp. 288: 'The Germanisation projects may seem visionary at first sight. Yet they were an organic element of both doctrine and blueprints which the Nazi leadership had adopted for the East ... a promotion of long-range, theologically conditioned products rather than an implementation of tasks that might have immediately contributed to the War effort'.
101. See p. 155 above, Koehl, *RKFDV*, pp. 146–51, 152–6, 224–7, Farquarson, *Plough & Swastika*, pp. 159–60 & 253–4.

NOTES TO CHAPTER EIGHT

1. R. Carson, *Silent Spring* (London, 1963). See P. Lowe and J. Goyder, *Environmental Groups in Politics* (London, 1983), and P. Kelly, *Fighting for Hope* (London, 1984).
2. For the German Youth Movement, see P. Stachura, *The German Youth Movement, 1900–45* (London, 1981) and W. Lacqueur, *Young Germany* (New York, 1962), Mosse, *Crisis, passim*, M. Kater, 'Die Artamanen', *HJ* (1971) 213, Jost Hermand, 'Meister Fidus; *Jugendstil*-Hippie to Aryan Faddist', *Comparative Literary Studies*, XII, 3 (1975). For the *Artamanen* magazine, *Die Kommenden*, and

the postcards, see BA NS/1285. R. Steiner, *Two Essays on Haeckel*, (repr. New York, 1935).

3. R. W. Darré (Carl Carlsson), 'Bauer und Technik', *Klüter Blätter*, 10 (1951). In reality, the average American farm was about 150 acres, and well within the *Erbhof* limit, see Hayami and Ruttan, *Agricultural Development; an International Perspective* (Baltimore, 1971).

4. Darré to Todt re; Seifert, 14.1.37 and 15.1.37, BA NS10/29. Nonetheless, Darré ordered hedgerows, copses and trees to be left as protection for wild life, see 24.1.40, BA NS10/37. For the Todt and Seifert compromise, see K. Ludwig's very good study, *Technik und Ingenieure im Dritten Reich* (Düsseldorf, 1974), p. 338.

5. H. Backe, *G.B.*, p. 20.

6. A. Seifert, 'Die Versteppung Deutschlands', *Deutsche Technik*, 4, (1936), and see Darré's Nachlass BA NL94II/1, 1a, for correspondence on organic farming. One paper sent to Darré by Seifert, not dated, argued that 'classical scientific farming' was a nineteenth-century liberal phenomenon, unsuited to the new era. It complained that imported artificial fertilisers, fodder and insecticides were poisonous in nature, and laid an extra burden on agriculture through transport costs. The writer called for a revolution in German agriculture towards a 'more peasant like, natural, simple method ... independent of capital ownership', but emphasised the need for a complete re-thinking of agricultural methods. 'The mere re-building of old peasant methods cannot help, because the internal interconnection of the old ways is gone;' these were typical arguments, then and now, BA NL94II/1. Darré kept a file of quotations from and comments on Steiner's works. BA NL94I/33.

7. Darré to Todt, op. cit.

8. A. Nisbett, *Konrad Lorenz* (London, 1976), pp. 77–91.

9. See R. Stauffer, 'Haeckel, Darwin and Ecology', *Quarterly Review of Biology*, xxxii (1957). The word does not seem to have acquired its normative sense until the mid-1930s.

10. Re-naming bio-dynamic farming, see Darré's speech to the National Peasant Council, 20.6.40, BA NS26/947.

> In 1933, my first task was to secure Germany against a blockade. It had to be as quick as possible, there was no time to worry about the right and wrong method of doing it ... I left the question of bio-dynamic farming open, and

gave it no publicity. But after the armistice with France, the danger of Germany's hunger receded, though a continental European blockade was still possible ... On June 18th, I visited a farm worked on bio-dynamic methods ... and established that this method is the best. The results speak clearly. If no explanation is forthcoming from scientists and previous agricultural teaching, that's their affair, the achievement and the result are ours. I have decided to support bio-dynamic farming ...

11. Darré to Peuckert, 4.8.41, BA NS35/9; letter from chemical company, G. Pacyna to Darré, 4.8.41, BA NL94II/1. On methane gas for peasant energy production, BA NL94I/27.

12. R. W. Darré, 'Der Lebensbaum unserer Altfarden [*sic*] im Lichte neuzeitlicher Naturwissenschaft', *Odal* (July 1935).

13. For Ludovici's articles, see BA NS26/945-6, and 'Skizze zur Gliederung der Bodenordnung', c.1936, in BA NS2/272, on the land as '*träger* of organic nature ... bound to landscape ... determining labour and economy'. Ludovici was representative for settlement (probably urban and rural) in Hess's office, see reports submitted to Schaub, a Hitler adjutant, BA NS10/53. Darré complained to Hess, 18.1.40, that Konrad Meyer had attacked 'bio-dynamic methods', BA NS10/37.

14. Heydrich to Darré, 18.10.41, BA NL94II/1. The Union of Anthroposophists was suspected of links with Freemasons, Jews and pacifists when it was closed in 1935, Haushofer, *Ideengeschichte*, p. 269.

15. Gauleiter Eggeling, formerly Keeper of the Seal, to Backe, re: Himmler, and enclosing copy letter to Himmler asking for support for the peasants, 1940, BA NL75/10, and see Riecke to Backe, 8.8.40, BA NL75/10, enclosing documents on organic farming. See also E. Georg, *Die Wirtschaftlichen Unternehmungen der SS* (Stuttgart, 1963), pp. 62–64, on Himmler's organic farms. Darré's successor as head of the SS RuSHA sent Himmler a list of farms which he visited in Poland after the German invasion, which would be useful for organic production, 20.11.39, BA NS2/60. Given the low level of articifial fertiliser use in Poland before the Second World War, it cannot have been difficult to find such places.

16. E.g., R. W. Darré, 'Bauerntum, Landarbeiter und Explosivegefahr', 10.3.34.

17. See here the works of Sir Albert Howard on compost and food values, and Sir George Stapledon on grass utilisation, and also G. V. Jacks and R. O. White, *The Rape of the Earth* (London, 1939).
18. Hellmut Bartsch, *Erinnerungen eines Landwirts, mit Erganzungen zu R. Steiner's Landwirtschaftlicher Impuls und seine Erfaltung* (Stuttgart, no date).
19. *Demeter*, 1939, p. 155. The poem was entitled, 'The Primacy of the Living'.
20. RNS meeting and Darré's visit, see Darré's diary, 13.5.40 and 18.6.40. He complained that 'Backe is always the same ... he thinks in terms of paper and statistics, and never organically, bio-dynamically'. Backe's widow comments that Backe was by no means hostile to the organic farming movement, but added revealingly that 'he hated the "mystic twilight"', interview with Frau Backe, July 1981. See also Haushofer, *Ideengeschichte*, p. 271, and p. 269 on Hess and the organic farming lobby. Darré sent Backe a pamphlet extolling the virtues of brown bread as early as February 1938, telling him that it was an important matter, with the health 'of the whole nation' involved, 1.2.38, BA NL94II/20.
21. E.g., *Landpost*, 13.9.40.
22. Questionnaire and replies to party leaders, 9.5.41, BA NL94II/1, and to RMEL Advisory Council, 19.5.41 in BA NS26/947. Darré wrote to Backe warning him of his forthcoming campaign in 1941, 7.6.41, BA NL94II/1.
23. Schacht to Darré, 30.5.41, BA NL94II/1. Frick, who had been exchanging letters with Dr Bartsch, was also sympathetic, 30.5.41, ibid. Goering to Darré, 20.6.41, criticised 'bio-dynamic propaganda' because it would lower market deliveries, and should wait until after the war. Darré replied confidently that many of the national leaders were 'positive' in their attitude, and further, that 'the Reichsführer SS [Himmler] in particular has built up agricultural farms for the SS via Pohl which are to be farmed in a bio-dynamic manner', 30.6.41, both BA NL94II/1. Criticism of organic farming methods came from Wagner, head of the civil administration in Alsace, who commented that many local farmers had moved over to organic farming, and had suffered losses in production of around 20% in the first year. 28.11.41, BA NL94II/1a.
24. Backe's warning to Darré, 3.7.41, BA NL94II/1a.
25. Darré to Heydrich re; nudists, 12.5.39, BA NL94II/49. Darré to Himmler and head of the SS Police, opposing the proceedings

against the Anthroposophists, 28.6.41, BA NL94II/1 and 1a: Darré
to Bormann, saying that 'members of the Anthroposophists Union
have nothing to reproach themselves with', 7.7.41, BA NL94II/1a.
H. S. Chamberlain was hostile to Anthroposophy, 'an offshoot of
Freemasonry and a danger to German values,' see Field, *Evangelist
of Race*, p. 508.

26. Heydrich to Darré, 18.10.41, BA NL94II/1. Darré, undeterred,
   appointed a working committee to study Steiner's works and
   organic farming, as part of his 'Friends of the German Peasantry
   Society'. They collected published and unpublished works by
   Steiner, BA NL94II/1d.
27. Peuckert, staff member of the Reichsamt für Agrarpolitik, to Rust,
   25.7.41, BA NL94II/1.
28. One top Anthroposophist, Dr R. Hauschka, gave evidence at Otto
   Ohlendorf's Nuremberg trial. He wrote that it was only
   Ohlendorf's intervention at a Hitler Conference that saved him,
   and four other Steiner men, from execution in 1941, an
   intervention that riled Himmler and Heydrich. Hauschka thinks
   that it was in revenge against this intervention that Ohlendorf was
   sent to Russia to head an *Einsatzgruppen* squad. Darré may have
   been lucky that the Allied assassination of Heydrich succeeded. See
   Dr R. Hauschka, *Wetterleuchten einer Zeitenwende* (Frankfurt, 1966),
   pp. 96–7, 101–2, and see also the memoirs of another
   Anthroposophist, Dr Wilhelm zur Linden, *Blick durches Prisma:
   Lebensbericht eines Arztes* (Frankfurt, 1966), pp. 111–2, on
   Ohlendorf.
29. Haushofer, *Ideengeschichte*, pp. 270–1.
30. BBC talk by Rolf Gardiner, on the *Europäische Bauernsendung*,
   September, 1940, transcript in author's possession, and Gardiner to
   Darré, 5.10.51, BA NL94II/20.
31. See Gardiner to Darré, on 'Kinship in Husbandry', 5.10.51,
   Gardiner to Darré, 25.11.51, and Gardiner to Reischle, 23.7.51, re-
   ference to Soil Association 'at the head of this movement', all BA
   NL94II/20. On Gardiner's estate and reafforestation, see J. Stewart
   Collis, *The Worm Forgives the Plough* (London, 1973), *passim*.
32. See Bartsch, *Erinnerungen*, and *Demeter* literature of the present day.
   The argument over organic farming and productivity is still
   continuing. The Oxford Farming Conference of 8.1.85, heard
   many speakers jeering at the idea that cheaper and healthier food
   could be produced by 'so-called organic methods'. The general
   secretary of the Society of Chemical Industry was particularly

scathing. The director of the Government's Agricultural and Advisory Service said there was a need for continuing scientific investigation, 9.1.85, *The Times*.

33. Dr H. Bartsch to Darré, 9.5.53, BA NL 94I/11.

NOTES TO CHAPTER NINE

1. Notice of Darré's sick leave, and notice that Backe was to take over his duties, both 20.5.42, BA R43II/1143. Darré's letter of resignation, 13.5.42, the formal appointment of Backe as Minister, 9.2.44, also BA R43II/1143. Himmler told Lammers to press Hitler to ensure Backe's appointment, 11.2.32, ibid. The annotated transcript of the Radio London broadcast by Dr Werner Clatt [*sic*] is in BA NL75/10-1. Dr Klatt was to be Darré's interrogator. Darré's diary, 14.3.42, refers to Backe's appointment, and see the ministry budget, in BA R2/17897, 1.8.42, which shows Darré as Minister, but Backe charged with carrying out all ministry business.

2. Darré's diary, 27.3.40. Cf Darré to Hitler, asking for sick leave, 14.6.37, BA NS10/35, and Darré to Lammers re; eczema, 1.11.40, and asthma, 27.10.43, in BA R43II/1143. Hitler sent him a 'get well soon' card, ibid.

3. See correspondence, file pages 74–80, between Willikens, Lammers and Darré, R43II/202a.

4. *TWCN*, vol xiv, p. 277.

5. Darré's diary, 22.4.45.

6. Report on agriculture, ED110/4, and in BA NL94I/26. Accounts of activities and people, and a history of the AD, BA NL94I/28. For his arrest, ED110/4-32. H. Kehrl, *Krisenmanager im Dritten Reich; Erinnerungen* (Düsseldorf, 1974), pp. 82–3. Erwin Goldmann, *Zwischen zwei Völkern; ein Rückblick* (Königswinter, 1975), pp. 192–3. See also, J. and D. Kimche, *The Secret Roads* (London, 1954).

7. Report on Darré's interrogation, by Dr Werner Klatt, April 1945, Wiener Library, N4f Darre J 36, and see pages above.

8. The Interrogator's report is undated, but is filed after the 1947 interrogations, see IfCH, Darré's interrogation, no. 1948/56.

9. Quoted in N. Bethell, *The Last Secret* (London, 1976), p. 248. Proposals by American academics and dignitaries to exterminate or just compulsorily sterilise all 80 million Germans were not only made but were discussed seriously. One English writer, Louis Nizer, in *What to do About Germany* (London, 1944), did, however, reject a sterilisation plan (three months for all German males, three years for all German women) put forward by a Harvard

anthropologist, but for practical reasons, not out of maudlin sentimentality, pp. 1–3. He also rejected plans to breed aggressiveness out of the Germans by forcible exogeny, and suggested compulsory labour service instead. One of Nizer's complaints was that some of the German patents seized by the US after the First World War had turned out to be defective: 'Never Again'. See also, ed. A. Weymouth, *Germany; Disease and Treatment*, published by the Parliamentary Policy Group (London, not dated), and Henry Morgenthau, Jr., *Germany is Our Problem* (USA, 1945) for similar answers. Alfredo de Zayas quotes a cheerful report by a Russian-born Sunday Times correspondent of the Russian army's attitude to mass rape. The correspondent quotes an officer as saying, 'They often raped old women of sixty ... seventy ... even eighty ... much to these grandmothers' surprise, if not downright delight', *Nemesis at Potsdam* (London, 1974), p. 69. The ignorance of one American counsel is commented on by M. Biddiss, 'The Nuremberg Trial: Two Exercises in Judgment', *JCH*, 16 (1981), p. 603.

10. Klatt, op. cit., p. 1. One official, I. Peter Dawes, noted on one of Darré's 1941 letters that Backe seemed not to have divulged much to Darré, BA NL94 II/20.

11. See Biddiss, op. cit. The archivist's introduction to the files of the Ministry of Agriculture at Coblenz mentions that barely a quarter of the files taken and stored by the Allies ever returned. Brandt, *Fortress Europe*, refers in his introduction to Ministry of Agriculture files in American private hands (1953).

12. See SS affidavit, copy in author's possession. Some 136,213 affidavits were filed for the SS defence, R. K. Woetzel, *The Nuremberg Trials in International Law* (London, 1960), p. 2.

13. John and Ann Tusa, *The Nuremberg Trial* (London, 1983), p. 93.

14. IfCH 1947 Interrogation, file pages 00062–3.

15. These twelve trials, held between 1946 and 1949, have been overshadowed by the trials of the top party leadership. The charges, prosecution and defence statements and judgements are published in *Trials of the War Criminals*, 1949 (*TWCN*), and should not be confused with the published IMT trial and its documents. The dissenting judgement by Judge Leon Powers, is in *TWCN*, xiv, pp. 871–940.

16. Letter to author from Dr Hans Merkel, 16.4.84, Darré's diary, 13.5.39.

17. See *TWCN*, xii–xiv, and Charter of the IMT, Articles 7, 8, 9, 10

and 16, published in *TWCN*, and also in Woetzel, *The Nuremberg Trials*, pp. 248–250.

18. *TWCN*, xiv, p. 871.
19. Although the legal and international status of the trials has been subject of considerable discussion, procedure has not. Translation errors are mentioned in the introduction to *TWCN*, i, pp. iii–iv. A careful reading of the material gives some idea of the problems. See e.g., the transcripts in W. Maser, *Nuremberg; A Nation on Trial* (London, 1977), Part II.
20. Observer, 30.5.37; P.R.O. MAF 39/20 plans drawn up in 1936 for the operation of the ministry in wartime.
21. For food movements during the war, see Brandt, *Fortress Europe*. See also Goering to Darré, February 1941, on the need to export bread to the Protectorate, Belgium and Norway, BA NL94II/20. Gross, *Polish Society*, pp. 103–4, 105–7, 109–13. The G-G was a food deficit country. See also Bannister, *I Lived under Hitler*, for a comparison of living standards in Poland and west Germany during the war.
22. See reference to the dismantling and seizure of German factories in A. Milward, *The Reconstruction of Western Europe, 1945–51* (London, 1984), p. 147, S. Siebel-Achenbach, 'The Social and Political Transformation of Lower Silesia, 1943–48', Draft Oxford D. Phil Thesis, and K. Belchling, 'The Nuremberg Judgments', in ed. W. E. Benton and G. Grimm, *Nuremberg: German Views of the War Trials* (Dallas, 1955), pp. 183–4. Not only were German prisoners-of-war retained in the USA after the war for labour service, but as many as possible were shipped out to America after the war for that purpose. See report of 2.1.47, C. E. King, Political Division, 'German Reactions to the Nuremberg Trials', quoted in Biddiss, op. cit., p. 613.
23. *TWCN*, xiv, 555–6. The judge's remark pre-dated by some twenty-seven years Britain's Race Relations Act, which has been seen as a totalitarian attack on free speech. The Judgment is in *TWCN*, xiv, pp. 416–862, including the judgments on the other defendants.
24. Op. cit., pp. 555–6.
25. Op. cit., p. 555.
26. Op. cit., p. 557. 'Unquestionably, the proceeds of the Aryanisation of farms and other Jewish property were in aid of and utilized in the program of rearmament and aggression'.
27. See e.g., notes on preparation of rations in Backe's papers, BA

NL75/12–1, and Ministry of Agriculture decree of 26.6.44, BA R43II/614.

28. *TWCN*, xiv, p. 560, and see chapter Four.
29. See discussion in Biddiss and Woetzel, op. cit.
30. Lutzhöft, *Nordischer Gedanke*, p. 23, but cf. Field, 'Nordic Racism', p. 536, and see chapter Two.
31. R. W. Darré, 'Walther Rathenau und das Problem des nordischen Menschen, *DE*, 7, 1926, and, ibid., 'Walther Rathenau und die Bedeutung der Rasse in der Weltgeschichte,' *DE*, 1928.
32. *Neuadel*, p. 230.
33. Ibid., p. 232.
34. R. W. Darré, 'Blut und Boden als Lebensgrundlage der nordischen Rasse', 22.6.30, reprinted *Blut und Boden* (Munich, 1941), pp. 17–29.
35. R. W. Darré, 'Das Schwein als Kriterium für nordische Völker und Semiten', *DE*, 3 (1927), and cf. R. Tannahill, *Food in History* (London, 1973), p. 64.
36. R. W. Darré, 'Damaschke und Marxismus' (MS, 1931), in BA NS26/949, and cf. published version in *VB*, Munich, 1.8.31, substantially cut.
37. D. Gasman, *The Scientific Origins of National Socialism* (New York, 1971), pp. 119–20.
38. R. W. Darré, 'Vom Friedenswillen der deutsche Bauern', autumn 1934, in BA R16/2047.
39. E.g., speech by Darré in 27.3.36, BA NL94II/1f.
40. See R. W. Darré, 'Stellungnahme des RBF zu der Frage, "RNS und Juden"', 10.3.35, BA R16/2057.
41. See E. D. Harrison, draft D. Phil Thesis, Oxford, 1982.
42. Correspondence between Backe and RNS officials, 1940–2, BA R14/266 and R14/267.
43. Darré, 'Stellungnahme'.
44. Darré, circular to RNS staff, 14.10.35, BA R16/2045.
45. Racial education, see Darré to a Norwegian nordicist, on need to produce picture books of Nordic nudes, 12.8.35, BA R16/2045. Letters about the error in *Neues Volk* are in BA NS2/74.
46. E.g., Darré's diary, 10.6.37, and Goldmann, op. cit.
47. Darré to Hitler, 11.12.39, BA NS10/37 and Darré's diary, 11.12.39. See also Darré's speech of 27.3.36, BA NL94 II/1f.
48. See de Zayas, *Nemesis at Potsdam*, *passim*.
49. *TWCN*, xiv, p. 824.
50. Op. cit., pp. 861–2.

51. The New York Times editorial; reported by the Jewish Telegraph Agency (August 27th, 1950) and, 18.8.50, cutting in Wiener Library files.
52. Darré to Dr Merkel, copy letter in author's possession, 31.10.50.
53. Ibid.
54. Darré to Lubbemaier, 26.5.53, BA NL94II/XI.
55. See correspondence in BA NL94I/20, and cf. R. Griffiths, *Fellow Travellers of the Right* (London, 1980), pp. 74–5ff.
56. Horthy and Darré, see Darré's minute in NDG/157, and letters from Hungarian veterans, ibid., letters from Horthy to Darré, BA NL941/II, and Horthy's *Memoirs* (Connecticut, 1957), p. 245.
57. Copy of Darré's notes at meeting in author's possession, 19–20. 12.52, p. 2.
58. In ED110/XI. Interview with H. Haushofer, 10.4.84.
59. *Goslarsche Zeitung*, cutting in ED110/XI. Darré asked to be buried in Goslar in his last will 'because here the European peasantry awoke and became self-aware'.
60. See Carmen Albert (née Darré), telegram to Alma and letter, thanking her for the roses, and describing the funeral, 10.9.53, ED110/XI.

Select Bibliography

SELECTED PUBLISHED WRITINGS OF DARRÉ

Many of Darré's articles were published first in journals, then as pamphlets, then in collected works. Articles referred to in this work which were reprinted in *Blut und Boden* have not been listed separately.

a. BOOKS

Das Bauerntum als Lebensquell der nordischen Rasse, Munich, 1929.
Erkenntnisse und Werden, Aufsätze aud der Zeit vor der Machtergreifung, ed. Marie Princess in Reuss zur Lippe, Goslar, 1940.
Neuadel aus Blut und Boden, Munich, 1930.
French Translation, *La Race*, tr. Melon, P., and Pfannstiel, A., Paris, 1939.
Um Blut und Boden: Reden und Aufsätze, ed. Hans Deetjens, Munich, 1941.

b. ARTICLES

'Bauerntum, Landarbeiterfragen, und Explosivgefahr, *Deutsche Zeitung* (10.4.34).
(pseud. Carl Carlsson), 'Bauer und Technik', *Klüter Blätter*, 10 (1951).
'Blut und Boden: ein Grundgedanke des Nationalsozialismus', *Odal* (1935), 794–805.
Damaschke, die Bodenreform und der Marxismus, MS 1931, Pamphlet, Munich, 1932, *VB* (1.8.31).
'Das Schwein als Kriterium für nordische Voelker und Semiten', *DE*, 3 (1927).
(pseud. Carl Carlsson), 'Die Ackerkrume ist etwas Lebendiges', *Goslarsche Zeitung* (5.4. 42).
'Gedanken zur Geschichte der Haustierwerdung', 1926, pr. 1940.
'Im Kampf für die Seele des deutschen Bauern', Pamphlet, Munich, 1935.
'Innere Kolonisation', *DE*, 4 (1926), 152–5.
'Landstand und Staat', *VB* (19–20.3.31).

'The National Food Estate', *Germany Speaks*, London, 1938, pp. 148–57.

'Walter Rathenau und das Problem des nordischen Menschen', and 'Walter Rathenau und die Bedeutung der Rasse in der Weltgeschichte', both *DE* 1926 and 1928 respectively, reprinted as one pamphlet, Munich, 1933.

'Zucht und Sitte; Die Neuordnung unserer Lebensgesetze', 1935, rep. 1942.

'Zur Förderung der Rassenhygiene', *Volk und Rasse*, 4 (1929).

SELECT BIBLIOGRAPHY

BOOKS

Abel, W., *Agrarpolitik*, Göttingen, 1967.

— *Agrarkrisen und Agrarkonjunktur, Berlin 1935*, repr. Hamburg, 1966.

Abraham, D., *The Collapse of the Weimar Republic*, Princeton, 1981.

Ackermann, J., *Heinrich Himmler als Ideologe*, Göttingen, 1970.

Ammon, Otto, *Die Gesellschaftsordnung und ihre natürlichen Grundlagen*, Jena, 1895.

Anon., *Die Leistungsfrage der Schweinehaltung* (undated, c. 1935), in series *Die Deutsche Zeugungsschlacht*.

— Report on Bessarabian Resettlement, MS, 1940.

Backe, H., 'Ziele und Aufgaben der Nationalsozialistischen Landvolkpolitik', Speech, 23 February, 1944, pub. Munich, 1944.

— *Das Ende des Liberalismus in der Wirtschaft*, Berlin, 1938.

— *Die Nährungsfreiheit Europas: Grossraum oder Weltwirtschaft?* Berlin, 1942.

Baker, J., *Race*, Oxford, 1974.

Balfour, Lady Eve, *The Living Soil*, London, 1941.

Bannister, S., *I Lived Under Hitler*, London, 1957.

Barkin, K., *The Controversy over German Industrialisation, 1890–1902*, Chicago, 1972.

Bartsch, H., *Erinnerungen eines Landwirts*, not dated, Stuttgart.

Bauer-Mengelberg, K., *Die Agrarfrage in Theorie und aktueller Politik*, Leipzig, 1931.

Bauer, Fischer, Lenz, *Human Heredity*, London, 1931.

Beck, J., Boehncke, H., and Vinnai, H., ed., *Leben im Faschismus: Terror und Hoffnung in Deutschland, 1933–45*, Hamburg, 1980.

Benton, W. E. and Grimm, G., ed., *Nuremberg: German Views of the War Trials*, Dallas, 1955.

Berghahn, V., *Der Stahlhelm: Bund der Frontsoldaten*, Düsseldorf, 1966.

Bergmann, Klaus, *Agrarromantik und Grosstadtfeindschaft*, Meisenheim am Glan, 1970.

Bertrand, R., *Le Corporatisme Agricole et L'Organisation des Marchés en Allemagne*, Paris, 1937.

— *Le National-Socialisme et La Terre*, Paris, 1938.

Bethell, N., *The War Hitler Won, September 1939*, London, 1972.

Bittermann, E., *Die Landwirtschaftliche Produktion in Deutschland, 1800–1950*, 1955. Sonderdruck aus Kühn-Archiv, Band 70, 1956.

Blackbourne, D., *Class, Religion and Local Politics in Wilhelmine Germany*, Yale, 1980.

Boberach, H., *Meldungen aus dem Reich*, Berlin, 1965.

Bollmus, R., *Das Amt Rosenberg und seine Gegner: Zum Machtkampf im national-sozialistischen Herrschaftssystem*, Stuttgart, 1970.

Bracher, K. D., *Die Deutsche Diktatur*, Berlin, 1969.

Brandt, K., *Management of Agriculture and Food in the German Occupied and other Areas of Fortress Europe: A Study in Military Government*, Stanford, 1953.

Brentano, L., *Agrarpolitik*, Stuttgart, 1897.

British Intelligence Objectives Sub-Committee Overall Report no. 6, *Some Agricultural Aspects in Germany during the Period 1939–45*, London, 1948.

Brobbeck, W., *Deutsche Getreidestatistik seit 1878*, Berlin, 1939.

Brozsat, M., *Nationalsozialistische Polenpolitik, 1939–45*, Stuttgart, 1961.

Buettner, J., *Der Weg zum Nationalsozialistische Recht*, Berlin, 1943: for texts of laws on land, credit, etc.

Caplan, A. L., *The Sociobiology Debate: Readings on the Ethical and Scientific Issues*, New York, 1978.

Cardinal, R., *German Romantics in Context*, London, 1975.

Carr, E. H., *German–Soviet Relations Between the Two World Wars, 1919–1939*, London, 1952.

Carsten, F. F., *The Reichswehr and Politics, 1918–1933*, Oxford, 1966, rev., 1973.

Cecil, R., *The Myth of the Master Race: Alfred Rosenberg and Nazi Ideology*, London, 1972.

Chase, A., *The Legacy of Malthus*, New York, 1977: includes material on Nazi eugenic ideas.

Chayanov, A., *The Theory of Peasant Economy*, ed., Thorner, D., et al., Illinois, 1966.

Control Council: Field Information Agency, Technical US Group, Control Council for Germany, *Summary of Report on German Resources and Technology in Agriculture*, London, 1945.

Coudenhove-Kalergi, R. N., *Adel*, Leipzig, 1922.

Dallin, A., *German Rule in Russia, 1941–45*, London, 1957.

Damaschke, A., *Aus Meinem Leben*, Leipzig, 1924.

— *Marxismus und Bodenreform*, Jena, 1923.

Darré, R. Oskar, *Meine Erziehung im Elternhause*, Wiesbaden, c. 1925.

David, E., *Sozialismus und Landwirtschaft*, Berlin, 1903.

Dawson, W., *The Evolution of Modern Germany*, London, 1908, repr. 1919.

Decken, H.v.d., *Der Deutsche Gemüsebau und seine Marktaussichten*, Berlin, 1949.

Department of Overseas Trade, *Economic Conditions in Germany*, 1934, 1936.

Deschner, G., *Heydrich: The Pursuit of Total Power*, London, 1981.

Dietze, C. von, *Preispolitik in der Welt Agrarkrise*, Berlin, 1936.

Eley, G., *Reshaping the Radical Right: Radical Nationalism and Political Change after Bismarck*, Yale, 1980.

Epstein, K., *The Genesis of German Conservatism*, Princeton, 1966.

Fallada, H., *Bauern, Bonzen und Bomben*, Berlin, 1931.

Farnsworth, Helen C., *Wartime Food Developments in Germany*, Stanford, 1942.

Farquharson, J. E., *The Plough and the Swastika: The NSDAP and Agriculture in Germany, 1928–1945*, London, 1976.

Field, G. C., *Evangelist of Race: the Germanic Vision of Houston Stewart Chamberlain*, Columbia, 1981.

Franz, G., *Quellen zur Geschichte des Deutschen Bauernstandes in der Neuzeit*, Munich, 1963.

Frauendorfer, S.v., and Haushofer, H., *Ideengeschichte der Agrarwirtschaft und Agrarpolitik im deutschen Sprachgebiet*, 2 vols; vol. 1, *Von den Anfängen bis zum 1. Weltkrieg*, Bonn, 1957.

Gasman, Daniel, *The Scientific Origins of National Socialism; Social Darwinism, Ernst Haeckel and the German Monist League*, New York/London, 1971.

Georg, Enno, *Die wirtschaftlichen Unternehmungen der SS*, Stuttgart, 1963.

Gerschrenkron, A., *Bread and Democracy in Germany*, Berkeley, 1943.

Gesell, S., *Die Neue Lehre vom Geld und Zins*, Berlin, 1911.

Gessner, D., *Agrarverbände in der Weimarer Republik*, Düsseldorf, 1976.

Glombowski, F., *Frontiers of Terror: the Fate of Schlageter and his Comrades*, London, 1935.

Görsler, A. and Troscher, T., *Energiewirtschaft und Maschinenverwendung im Siedlerbetrieb*, Berlin, 1933.

Goldmann, E., *Zwischen Zwei Völkern: ein Rückblick. Erlebnisse und Erkenntnisse*, Königswinter, 1975.

Gollwitzer, H., ed., *Europäische Bauernparteien im 20 Jahrhundert*, Stuttgart, 1977.

Greenwood, H., *The German Revolution*, London, 1934.

Griffiths, R., *Fellow Travellers of the Right: British Enthusiasts for Nazi Germany*, London, 1980.

Gritzbach, E., *Hermann Goering: The Man and his Work*, London, 1938.

Gross, Jan T., *Polish Society under German Occupation: The General-Gouvernement 1939-44*, Princeton, 1979.

Grundmann, F., *Agrarpolitik im Dritten Reich; Anspruch und Wirklichkeit des Reichserbhofgesetzes*, Hamburg, 1979.

Günther, H. F.-K., *Kleine Rassenkunde des deutschen Volkes*, Munich, 1922.

— *Die Verstädterung: Ihre Gefahren für Volk und Staat*, Leipzig, 1934.

— *The Racial Elements of European History*, London, 1927.

Guillebaud, C. W., *The Economic Recovery of Germany, 1933-1938*, London, 1939.

Haase, L., *Aufstand in Niedersachsen*, mim. 1942.

Haeckel, E., *The Wonders of Life*, London, 1905.

Hagen, W. W., *Germans, Poles and Jews: Nationality Conflict in the Prussian East, 1772-1914*, Chicago, 1979 (predominantly concerned with German anti-Polish movements).

Hamilton, R., *Who Voted for Hitler?* London, 1982.

Hanau, A. and Plate, R., *Die deutsche landwirtschaftliche Preis-und Marktpolitik im Zweiten Weltkrieg*, Stuttgart, 1975.

Hassell, U. v., *Diaries, 1938-44*, Eng. tr. London, 1948.

Hauner, M., *India in Axis Strategy*, London, 1980.

Haushofer, H., *Bauern Fuhren den Pflug nach Osten: Wie des Reiches älteste Ostmark Entstand*, undated, probably 1939-40.

— *Bauern zwischen Ost und West*, Stuttgart, 1946.

— *Ideengeschichte der Agrarwirtschaft und Agrarpolitik im deutschen Sprachgebiet*, vol. 2, *Vom Ersten Weltkrieg bis zur Gegenwart*, Bonn, 1958.

— *Die deutsche Landwirtschaft im Technischen Zeitalter*, Stuttgart, 1963.

— *Mein Leben als Agrarier*, Munich, 1982.

Hayami, Y. and Ruttan, V., *Agricultural Developemt: an International Perspective*, Baltimore, 1971.

Heberle, R., *Landbevölkerung und Nationalsozialismus. Eine soziologische*

Untersuchung der politischen Willensbildung in Schleswig-Holstein 1918 bis 1932, Stuttgart, 1963.

Heidegger, M., *The End of Philosophy*, London, 1956.

Henning, F.-W., *Landwirtschaft und ländliche Gesellschaft in Deutschland, 1750–1976*, vol. 2, Paderborn, 1978.

Hertz-Eichenrode, D., *Politik und Landwirtschaft in Ostpreussen, 1919–30*, Cologne-Opladen, 1969.

Herzl, T., *Der Judenstaat*, Vienna, 1905.

Hinsley, F. H., *British Intelligence in the Second World War: its Influence on Strategy and Operations*, London, 1981, pp. 133, 286, 426–7, 540, Ministry of Economic Warfare reports on the German food supply situation.

Hinton-Thomas, R., *Nietzsche in German Politics and Society, 1890–1918*, Manchester, 1984.

Hirschfield, G. and Kettenacker, L., *Der 'Führer-Staat': Mythos und Realität*, London, 1980.

Hitler, A., *Mein Kampf*, tr. J. Murphy, London, 1939.

— *Speeches*, vol. 1, Domestic, tr. Norman Baynes, New York, 1942.

— *Hitler's Secret Book*, New York, 1961.

— *Reden*, vol. 1, ed. Domarus, M., Würzburg, 1962.

HMSO, *How Britain was Fed in Wartime: Food Control, 1939–45*, London, 1946.

Höhne, H., *The Order of the Death's Head*, New York, 1970.

Hohenstein, A., *Wartheländisches Tagebuch aus den Jahren 1941–2*, Stuttgart, 1961.

Holt, J. B., *German Agricultural Policy, 1918–1934: The Development of a National Philosophy towards Agriculture in Post-War Germany*, Chapel Hill, 1936.

Hussain, A., in Hussain, A. and Tribe, K., ed., *Marxism and the Agrarian Question*, 2 vols.; vol. 1, *German Social Democracy and the Peasantry 1890–1907*, London, 1981.

International Labour Organisation, *Rural Exodus in Germany*, Studies and Reports, no. 12, Geneva, 1935.

Ipsen, G., *Das Landvolk: Ein soziologischer Versuch*, Hamburg, 1933.

Irving, D., *Hitler's War*, London, 1979.

— *The War Path*, London, 1979.

Jacks, G. V. & Whyte, R. O., *The Rape of the Earth*, London, 1939.

Jasny, N., *Bevölkerungsrückgang und Landwirtschaft*, Berlin, 1931.

Kater, M. H., *Das 'Ahnenerbe' der SS, 1935–45*, Stuttgart, 1974.

Kautsky, K., *Die Agrarfrage; Eine Ubersicht über die Tendenzen der Modernen Landwirtschaft und der Agrarpolitik der Sozialdemokratie*, Stuttgart, 1899.

— *Die Sozialisierung der Landwirtschaft*, Berlin, 1919.

Kehrl, H., *Krisenmanager im Dritten Reich: Erinnerungen*, Düsseldorf, 1973.

Kele, M., *Nazis and Workers*, Chapel Hill, 1972.

Kern, F., *Stammbaum und Artbild der deutschen Bauern und ihrer Verwandten*, Bonn, 1927.

Kelly, A., *The Descent of Darwin: the Popularisation of Darwinism in Germany, 1860–1914*, Chapel Hill, 1981.

Kershaw, I., *Der Hitler–Mythos: Volksmeinung und Propaganda im Dritten Reich*, Stuttgart, 1980.

Kimche Bros., *The Secret Roads*, London, 1954.

Klemperer, K. von, *Germany's New Conservatism*, Princeton, 1957.

Knickerbocker, H. R., *Germany: Fascist or Soviet?* London, 1932.

Koch, W. H-J., *Der Sozialdarwinismus, seine Entstehung und sein Einfluss auf das imperialistische Denken*, Munich, 1973.

Koehl, R. L., *The Black Corps; the Structure and Power Struggles of the Nazi SS*, Wisconsin, 1983.

— *RKFDV: German Resettlement and Population Policy*, Cambridge, Mass., 1957.

Krausnick, H., and Wilhelm, H. H., *Die Truppe des Weltanschauungskrieges. Die Einsatzgruppen der Sicherheitspolizei und des SD*, Stuttgart, 1980.

Krieger, L., *The German Idea of Freedom*, Chicago, 1972.

Krochow, C. von, *Die Entscheidung: eine Untersuchung über Ernst Jünger, Carl Schmitt, Martin Heidegger*, Stuttgart, 1958.

Kröger, E., *Der Auszug aus der alten Heimat. Die Umsiedlung der Baltendeutschen*, Tübingen, 1967.

Kulischer, E., *The Displacement of Population in Europe*, ILO Studies and Reports, Series O, Migrations, no. 8, Montreal, 1943.

Lacqueur, W., *Young Germany*, New York, 1962.

— *Russia and Germany; a Century of Conflict*, London, 1965.

Lagarde, Paul de, *Deutsche Schriften*, Göttingen, 1937 (1878).

Lane, B. M., *Architecture and Politics in Nazi Germany, 1918–1945*, Cambridge, Mass., 1968.

— and Rupp, L., *Nazi Ideology Before 1933: a Documentation*, Manchester, 1978.

Lang, Jochen von, *The Secretary: Martin Bormann, the Man Who Manipulated Hitler*, New York, 1979.

Lehmann, H. G., *Zur Agrarfrage in der Theorie und Praxis der deutschen und internationalen Sozialdemokratie. Vom Marxismus zum Revisionismus und Bolschevismus*, Tübingen, 1970.

Leopold, J. A., *Alfred Hugenberg: The Radical Nationalist Campaign Against the Weimar Republic*, Yale, 1978.

Lindberg, J., *Food, Famine and Relief*, League of Nations, Geneva, 1940.

Lipset, S., *Agrarian Socialism*, California, 1967.

Löber, D. A., *Diktierte Option. Die Umsiedlung der Deutsch–Balten aus Estland und Lettland, 1939–41*, Neumunster, 1972.

Lougee, R., *Paul de Lagarde, a Study of Radical Conservatism in Germany*, Cambridge, Mass., 1962.

Ludwig, K. H., *Technik und Ingenieure im Dritten Reich*, Düsseldorf, 1974.

Lutzhöft, H-J., *Der Nordische Gedanke in Deutschland, 1920–40*, Stuttgart, 1970.

Manuel, F. E. and F. P., *Utopian Thought in the Western World*, Oxford, 1979.

Manvell, R. and Fraenkel, H., *Himmler*, London, 1965.

Maschmann, M., *Fazit. Kein Rechtfertigungsversuch*, Stuttgart, 1963.

Maser, W., *Nuremberg. A Nation on Trial*, London, 1977.

Mason, T. W., *Sozialpolitik im Dritten Reich: Arbeiterklasse und Volksgemeinschaft*, Opladen, 1977.

Massingham, H. J., ed., *The Small Farmer*, London, 1947.

Mayhew, A., *Rural Settlement and Farming in Germany*, London, 1973.

Mazière, C. de la, *Ashes of Honour*, tr. Francis Stuart, London, 1976.

Mehrens, B., *Die Marktordnung des RNS*, Berlin, 1938.

Meisner, A., *Die Agrarverfassung der Deutschen im Sudetenland und den Westkarpathen*, unbound, probably Germany, c. 1931.

Merkl, P., *Political Violence under the Swastika. 581 Early Nazis*, Princeton, 1976.

— *The Making of a Stormtrooper*, Princeton, 1980.

Meyer, Konrad, 'Lebensbericht', unpub. MS.

— ed., *Gefüge und Ordnung der deutschen Landwirtschaft als Gemeinschaftsarbeit des Forschungsdienstes*, Berlin, 1939.

Milward, A., *The Reconstruction of Western Europe, 1945–1951*, London, 1984.

— *War, Economy and Society*, 1939–45, London, 1977.

Mitrany, D., *Marx Against the Peasant. A Study in Social Dogmatism*, New York, 1961.

Moeller v.d. Bruck, *The Third Reich*, Berlin, 1923.

Morrow, I., *The Peace Settlement in the German–Polish Borderlands: A Study of Conditions Today in the Pre-War Prussian Provinces of East and West Prussia*, London, 1936.

Mosse, G. L., *The Crisis of German Ideology*, London, 1964.

— *Nazi Culture, Intellectual, Cultural and Social Life in the Third Reich*, London, 1966.

— *Germans and Jews: The Right, the Left and the Search for a Third Force in pre-Nazi Germany*, New York, 1971.

Müller-Sternberg, R. and Nellner, W., *Deutsche Ostsiedlung: eine Bilanz für Europa*, Bielefeld, 1969.

Murray, K., *History of Agriculture in Britain in the Second World War*, London, 1955.

Nisbett, A., *Konrad Lorenz*, London, 1976.

Noakes, J., *The Nazi Party in Lower Saxony, 1921-33*, Oxford, 1971.

— and Pridham, G., *Documents on Naziism, 1919-1945*, London, 1974.

Nolte, E., *The Three Faces of Fascism*, tr. L. Vennevitz, U.S.A., 1969.

Orlow, D., *A History of the Nazi Party, 1919-1933*, Pittsburgh, 1969.

Payne, S. G., *Fascism: Comparisons and Definitions*, London, 1980.

Peterson, E., *The Limits of Hitler's Power*, Princeton, 1969.

Petzina, D., *Autarkiepolitik im Dritten Reich*, Stuttgart, 1968.

Polonsky, A., *The Little Dictators*, London, 1975.

Poole, K., *Germany's Financial Policies*, Harvard, 1939.

Proudfoot, M., *European Refugees*, 1939-52, London, 1957.

Puhle, H. J., *Politische Agrarbewegungen in kapitalistischen Industriegesellschaften*, Göttingen, 1975.

Pulzer, P., *The Rise of Political Anti-Semitism in Germany and Austria*, New York, 1964.

Rauschning, H., *The Makers of Destruction*, London, 1942.

Reill, P. H., *The German Enlightenment and the Rise of Historicism*, California, 1975.

Reinhardt, F., *Wie schlägt man die Erzeugungsschlacht*, Essen, 1935.

Reischle, H., *RBF Darré, der Kämpfer um Blut und Boden*, Berlin, 1933.

— and Saure, W., *Der RNS, Aufbau, Aufgaben, Bedeutung*, Berlin, 1934.

Rhode, G., ed., *Die Ostgebiete des Deutschen Reiches*, Würzburg, 1955.

Rich, N., *Hitler's War Aims*, 2 vols, London, 1973-4.

Royal Institute of International Affairs, *World Agriculture: An International Survey*, London, 1932.

Ruhland, G., *The Ruin of the World's Agriculture and Trade*, London, 1896.

— *System der Politischen Oekonomie*, 3 vols, Berlin, 1903-8. (Note: Many British catalogues spell Ruhland's name with an umlaut over the 'u', and *Ruin* is printed throughout in this way.)

Salomon, E. von, *Der Fragebogen*, Hamburg, 1951.

Schieder, T., ed., *Documents on the Expulsion of the Germans from Eastern-Central Europe*, 4 vols, Vol. 2-3, Göttingen 1951, abridged English version, 1961.

Schilke, F., *Trakhener Horses Then and Now*, Oklahoma, 1977.

Schmokel, W. W., *Dream of Empire: German Colonialism, 1919–1945*, London, 1964.

Schoenbaum, D., *Hitler's Social Revolution: Class and Status in Nazi Germany*, London, 1969.

Schoenberg, H. W., *Germans from the East. A Study of their Migration, Resettlement and Subsequent Group History since 1945*, Hague, 1970.

Schöpke, K., *Der Ruf der Erde: Deutsche Siedlung in Vergangenheit und Gegenwart*, Leipzig and Berlin, 1935.

— *Deutsche Ostsiedlung*, Berlin, 1943.

— *Deutsche Ostsiedlung in Mittelalter und Neuzeit*, Cologne, 1971.

Schohl, H., *Die Grundlagen und Wandlungen der Ost Landwirtschaft in der Nachkriegszeit*, 1937.

Schultze, P., *Die Aufgabengebiete der Erzeugungsschlacht*, Berlin, 1936.

Schultze-Naumburg, P., *Kunst aus Blut und Boden*, Leipzig, 1934.

Schweitzer, A., *Big Business in the Third Reich*, Bloomington, 1964.

Schwerin-Krosigk, L., *Es Geschah in Deutschland*, Tübingen and Stuttgart, 1951.

Seraphim, H. J., *Deutsche Bauern-und Landwirtschaftspolitik*, Leipzig, 1939.

Sering, M., *Deutsche Agrarpolitik auf geschichtlicher und landeskundiger Grundlage*, Leipzig, 1934.

— *Kritik vom Erbhofgesetz*, DAP, 1934.

— and Dietze, C. von, *Die Vererbung des ländlichen Grundbesitzes in der Nachkriegszeit*, Munich and Leipzig, 1930.

— ed., *Agrarverfassung der deutschen Auslandssiedlungen in Osteuropa*, Berlin, 1939.

Smelser, R. M., *The Sudeten Problem, 1933–1938; Volkstumspolitik and the Formulation and Nazi Foreign Policy*, Folkestone, 1975.

Snyder, C., *Soldiers of Destruction: The History of the Death's Head Battalion 1933–45*, Princeton, 1977.

Sohn-Rethel, Alfred, *Ökonomie und Klassenstruktur des deutschen Faschismus; Aufzeichnung und Analysen*, Frankfurt, 1973.

Sombart, W., *Die deutsche Volkswirtschaft im 19. und am Anfang des 20. Jahrhundert*, Berlin, 1927.

— *A New Social Philosophy*, tr. K. F. Geiser, Princeton, 1937.

Speer, A., *Inside the Third Reich*, London, 1970.

— *The Slave State: Heinrich Himmler's Master Plan for SS Supremacy*, London, 1981.

Spengler, O., *Decline of the West*, tr. Atkinson, C. F., London, 1932.

Stachura, P. D., ed., *The Shaping of the Nazi State*, London, 1978.

— *The German Youth Movement 1900–45*, London, 1981.

— *Gregor Strasser and the Rise of Nazism*, London, 1983.

Stark, G. D., *Entrepreneurs of Ideology: Neoconservative Publishers in Germany, 1890–1933*, Chapel Hill, 1981.

Statistisches Handbuch für die Provinz Ostpreussen, 1938.

Steiner, J. M., *Power Politics and Social Change in National Socialist Germany*, The Hague, 1975.

Steiner, R., *Two Essays on Haeckel*, New York, 1936.

Stern, Fritz, *The Politics of Cultural Despair*, Berkeley, 1961.

Stern, J. P., *Ernst Jünger: A Writer of our Time*, London, 1952.

Stewart, J. Collis, *The Worm Forgives the Plough*, London, 1947, repr., 1973.

Strub, H., *Das deutsche landwirtschaftliche Genossenschaftswesen in RNS*, Berlin, 1937.

Struve, W., *Élites Against Democracy: Leadership Ideals in Bourgeois Political Thought in Germany, 1890–1933*, Princeton, 1973.

Tannahill, R., *Food in History*, London, 1973.

Taylor, R., *Literature and Society in Germany, 1918–1945*, Brighton, 1980.

Thomas, G., *Geschichte der deutschen Wehr-und Rüstungswirtschaft, 1918–1943/5*, ed., Birkenfeld, W., Coblenz, 1966.

Thünen, J. von, *The Isolated State*, ed. Hall, P., London, 1966.

Tocqueville, A. de., *The European Revolution and Correspondence with Gobineau*, ed. Lukacs, J., 1843–59, New York, 1959.

Tornow, W., *Chronik der Agrarpolitik und Agrarwirtschaft des deutschen Reiches, 1933–45*, Hamburg, 1972.

Tracy, M., *Agriculture in Western Europe since 1880*, London, 1964.

Trials of the War Criminals before the Nuremberg Military Tribunal, vols 4, 13, 14, Washington, 1949.

Trittel, G. J., *Die Bodenreform in der Britischen Zone 1945–59*, Stuttgart, 1975.

Turner, H. J., ed., *Naziism and the Third Reich*, New York, 1972.

Tusa, J. & A., *The Nuremberg Trial*, London, 1983.

Vogelsang, R., *Der Freundeskreis Himmlers*, Göttingen, 1972.

Wagener, Otto, *Hitler aus Nächster Nähe*, ed. Turner, H. A., Frankfurt a. Main, 1978.

Waite, R., *'Vanguard of Naziism': The Free Corps Movement in Postwar Germany, 1918–1923*, Cambridge, Mass., 1952.

Warriner, D., *The Economics of Peasant Farming*, London, 1939–1964.

Weber, A., *Productivity Growth in German Agriculture, 1850–1970*, Staff Paper for the Dept of Agriculture and Applied Economics, University of Minnesota, 1973.

Weizsäcker, E. von, *Memoirs*, Chicago, 1951.

Welch, D., *Propaganda and the German Cinema*, Oxford, 1983.

Willikens, W., *Nationalsozialistische Agrarpolitik*, Munich, 1931 (preface by Darré).

Woetzel, R. K., *The Nuremberg Trials in International Law*, London, 1960.

Wunderlich, F., *Farm Labour in Germany, 1810–1945*, Princeton, 1961, esp. Part III.

Yates, P. Lamartine, *Food Production in Western Europe*, London, 1940.

Zayas, A. de, *Nemesis at Potsdam*, London, 1979.

JOURNALS AND UNPUBLISHED DISSERTATIONS

Angress, W. T., and Smith, B. F., 'Diaries of Heinrich Himmler's Early Years', *JMH*, 31 (1959).

Balogh, T., 'The National Economy of Germany', *Economic Journal* (Sept. 1938).

Bessel, R., 'East Germany as a Structural Problem in the Weimar Republic', *Social History*, 3 (1978), 210–3.

— 'The SA in the Eastern Regions of Germany, 1925–37', Oxford D. Phil. Thesis, 1980.

Biddiss, M., 'The Nuremberg Trial: Two Exercises in Judgment', *JCH*, 16 (1981), 579–615.

Braatz, W. E., 'Two Neo-Conservative Myths in Germany 1919–32: The "Third Reich" and "New State" ', *JHI*, 32 (1971), 569–84.

Bramwell, A. C., 'National Socialist Agrarian Theory and Practice, with Special Reference to Darré and the Settlement Movement', Oxford D. Phil Thesis, 1982.

— 'German Identity Transformed', *JASO*, XVI/1 (1985).

— 'R. Walther Darré; Was this man Father of the Greens?' *History Today*, 34 (Sept., 1984).

— 'Small Farm Productivity under the Nazis', *Oxford Agrarian Studies*, XIII (1984).

Brandt, K., 'Junkers to the Fore Again', *Foreign Affairs* (1935), 120–34.

— 'German Agricultural Policy: Some Lessons', *Journal of Farm Economics*, 19 (1937).

Brannig, R., 'Die Leistungsfähigkeit des Siedlungsbetriebes im Vergleich zum Grossbetriebes', *BuL* Sonderheft, 98 (1934).

Bruchhold-Wahl, H., 'Ökonomische Situation der Ostelbischen Grossgrundesitzer am Ende des 19 Jahrhunderts', 'UEA', 1979.

Clauss, W., 'Erfahrungen aus 50 Jahren Agrarpolitik', Schleswig-Holstein *Bauern-Verband*, 1979.

Decken, Hans v.d., 'Entwicklung der Selbstversorgung Deutschlands mit landwirtschaftlichen Erzeugnissen', *BuL*, 138 (1938).

— 'Die Mechanisierung in der Landwirtschaft', *Vjhz Wirtschaftsforschung* 3 (1938-9).

Durant, J., 'The Meaning of Evolution; Post-Darwinian Debates on the Significance for Man of the Theory of Evolution, 1858-1908', Cambridge, Ph.D. Thesis, 1977.

Eisenblatter, G., 'Grundlinien der Politik des Reiches gegenüber dem General Gouvernement, 1939-45', Frankfurt a. Main Diss., 1969.

Eley, G., 'Nationalism and Social History', *Social History*, 6 (1981), 83-109.

Farquharson, J., 'The NSDAP and Agriculture in Germany, 1928-38', Canterbury, Ph.D. Thesis, 1972.

Farr, I., 'Populism in the Countryside: the Peasant Leagues in Bavaria in the 1890s', in, ed., Evans, R., *Society and Politics in the Wilhelmine Era*, London, 1979.

Fiederlein, F. M., 'Der deutsche Osten und die Regierungen Brüning, Papen, Schleicher', Würzburg Phil. Diss., 1966.

Field, G. G., 'Nordic Racism', *JHI*, 38 (1977), 523-40.

Fraenkel, E., 'German-Russian Relations since 1918', *Review of Politics*, 2 (1940).

Frölich, E., 'Politische und Soziale Macht auf dem Lande', *VjhfZg*, 25 (1977).

Galbraith, J. K., 'Hereditary Land in the Third Reich', *Quarterly Journal of Economics*, 53 (1938-9), 465-76.

Gessner, D., 'Agricultural Protectionism in the Weimar Republic', *JCH*, 12 (1977), 759-78.

— 'The Dilemma of German Agriculture during the Weimar Republic', in, ed., Bessel, R. and Feuchtwanger, E. J., *Social Change and Political Developmemt in Weimar Germany*, London, 1981.

Gies, H., 'R. Walther Darré und die nationalsozialistische Bauernpolitik in den Jahren 1930 bis 1933', Frankfurt a. Main, Phil. Diss., 1966.

— 'NSDAP und landwirtschaftliche Organisation in der Endphase der Weimarer Republik', *VjhfZg* (1967).

— 'Der RNS—Organ Berufsständischer Selbstverwaltung oder Instrument Staatlicher Wirtschaftslenkung', *ZAGAS*, 21 (1973), 216-33.

— 'Die Rolle des RNSt im nationalsozialistischen Herrschaftssystem', in, ed., Kettenacker, L. et al., *Der 'Führer-Staat', Mythos und Realität*, Stuttgart, 1981.

Gossweiler, K. and Schlicht, A., 'Junker und NSDAP', *Zeitschrift für Geschichtswissenschaft*, 15 (1967), 644-60.

Graham, Loren R., 'Science and Values: the Eugenics Movement in Germany and Russia in the 1920s', *AHR*, 82 (1977), 1133-64.

Haag, J., 'Othmar Spann and the Politics of Totality: Corporatism in Theory and Practice', Rice University, Ph.D. Diss., 1969.

Halliday, R. J., 'Social Darwinism: a Definition', *Victorian Studies*, 14 (1971), 389-406.

Haushofer, H., 'Die Idealvorstellung von deutsche Bauern', *ZAGAS*, 26 (1978).

Heiber, Helmut, Intro. to 'Der Generalplan Ost' (Docs), *VjhfZg*, 6 (1958), 281-324.

Heinemann, J., 'Constantin von Neurath and German Policy at the London Economic Conference of 1933: the Background to the Resignation of Alfred Hugenberg', *JMH*, 41 (1969).

Herf. J., 'Reactionary Modernists in Weimar and Nazi Germany', *JCH*, 19/4, Oct., 1984.

Hermand, Jost, 'Nazi Concepts of Matriarchy, *JCH*, 19/4, Oct., 1984.

— 'Meister Fidus: *Jugenstil*-Hippie to Aryan Faddist', *CLS*, XII (3) 1975.

Hermann, A. R., 'Erbhof und Kredit', *Zeitschrift für die Gesamte Staatswissenschaft*, 95 (1935).

Hillgruber, Andreas, 'Die Endlösung und das deutsche Ostimperium als Kernstück des rassenideologischen Programms des Nationalsozialismus', *VjhfZg*, 20 (1972).

Hörnigk, R., 'Germany', in *Year-Book of Agricultural Cooperation*, 1937.

Hollman, A. H., 'Die Agrarkrise der ost-und südosteuropäischen Staaten', *DAP* (1932).

Holmes, K. R., 'The Forsaken Past: Agrarian Conservatism and National Socialism in Germany', *JCH*, 174, Oct., 1982.

Holt, N. R., 'Ernst Haeckel's Monistic Religion', *JHI*, 32, 1971.

Hunt, J. C., 'The "Egalitarianism" of the Right; the Agrarian League in South-West Germany 1893-1914', *JCH*, 10 (1975), 513-31.

Jasny, Marie, 'Some Aspects of German Agricultural Settlement', *PSQ*, 2 (1937), 208-40.

Kater, M. H., 'Die Artamanen: Völkische Jugend in der Weimarer Republik', *HZ*, 213 (1971).

Kershaw, I., 'The Führer Image and Political Integration', in, ed., Kettenacker, L., *Der 'Führer Staat'*, *Mythos und Realität*, Stuttgart, 1981.

Kitani, T., 'Brünings Siedlungspolitik und sein Sturz', *ZAGAS*, 14 (1966), 54-82.

Kluke, P., 'Nationalsozialistische Europaideologie', *VjhfZg*, 3 (1955), 240-75.

Knox, McGregor, 'Fascist Italy and Nazi Germany', *JMH*, June, 1984.

Koch, H. W., 'Hitler and the Origins of the Second World War; Second Thoughts on the Status of some of the Documents', *Historical Journal*, 11 (1968), 125–44.

Köhler, H., 'Arbeitsbeschaffung, Siedlung und Reparationen in der Schlussphase der Regierung Brüning', *VjhfZg*, 17 (1969), 270–307.

Kozauer, N. J., 'The Carpatho-Ukraine between the Two World Wars, with Special Emphasis on the German Population', Rutgers University Ph.D. Thesis, 1964.

Krüdener, J. von, 'Zielkonflikt in der nationalsozialistischen Agrarpolitik, ein Beitrag zur Diskussion des Leistungsproblems in zentral gelenkten Wirtschaftssystemen', *ZWS*, 4 (1974), 335–62.

Kühnl, R., 'Zur Programmatik der nationalsozialistischen Linken: Das Strasser-Programm von 1925–6', *VjhfZg* (1966).

Kulischer, E., 'Population Transfer', *South Atlantic Quarterly* (1946).

Kummer, K., 'Die Entwicklung der landwirtschaftlichen Siedlung in der Provinz Grenzmark Posen-Westpreussen', Berlin, Diss., 1919.

Lange, O., 'Gustav Ruhland; System der Politischen Oekonomie', *Odal* (1935).

Lavelaye, Emil de, 'Land System of Belgium and Holland', in *Systems of Land Tenure in Various Countries*, London, 1870.

Lee, J. J., 'Labour in German Industrialisation', in *Cambridge Economic History of Europe*, vol. 7, Cambridge, 1978.

Lehmann, J., 'Zum Zusammenbruch der Kriegsernährungswirtschaft im faschistischen Deutschland, 1944–5', in *Probleme der Agrargeschichte des Feudalismus und Kapitalismus*, Rostock, 1978.

— 'Untersuchungen zur Agrarpolitik und Landwirtschaft im faschistischen Deutschland während des Zweiten Weltkrieges, 1942–5', Rostock, Diss., 1978.

— 'Faschistische Agrarpolitik im Zweiten Weltkrieg; zur Konzeption von Herbert Backe', *Zeitschrift für Geschichtswissenschaft*, 19 (1980).

Lenz, W., 'Erbhöfe für baltische Restgutbesitzer im Warthegau', *JBD*, 29 (1982).

Levine, H., 'Local Authority and the SS-State; the Conflict over Population Policy in Danzig West-Prussia, 1939–45', *CEH*, 2 (1969).

Lewis, G., 'The Peasantry; Rural Change and Conservative Agrarianism. Lower Austria and the Turn of the Century', *Past and Present*, 80 (1978).

Loewenberg, P., 'The Adolescence of Heinrich Himmler', *AHR*, 76 (1971), 612–41.

Loomis, C. P. and Beegle, J. A., 'The Spread of German Naziism in Rural Areas', *ASR*, 11 (1946), 724–34.

Lovin, C. R., 'German Agricultural Policy 1933–36', University of North Carolina Ph.D. Thesis, 1965.

— '"Blut und Boden": The Ideological Basis of the Nazi Agricultural System', *JHI*, 28 (1967), 279–88.

— 'Die Erzeugungsschlacht 1934–6', *ZAGAS*, 22 (1974), 209–20.

Mason, T. W., 'National Socialist Policies Towards the Working Classes 1925–39', Oxford D.Phil. Thesis, 1971.

Meinhold, W., 'Gustav Ruhlands Mittelstandtheorie', *Odal* (1935).

Milward, A., 'German Economic Policy towards France, 1942–44', in, ed. Barne, K., *Studies in International History*, London, 1967.

— 'French Labour and the German Economy', *Economic History Review*, 23 (1970).

— 'Fascists, Nazis and Historical Method', *History* (1982).

Muehlberger, D., 'The Sociology of the NSDAP: the Question of Working-Class Membership', *JCH*, 15 (1980).

Noakes, J., 'Conflict and Development in the NSDAP, 1924–27', *JCH*, 1 (1966).

— 'Nazism and Eugenics: the Background to the Nazi Sterilization Law of July, 1933', in, ed. R. Buller, et. al., *Ideas Into Politics*, London, 1984.

O'Lessker, Karl, 'Who Voted for Hitler? A New Look at the Class Basis of Nazism', *AJS*, 74 (1968).

Passchier, N., 'The Electoral Geography of the Nazi Landslide', in, ed. Stein, E. L., et al., *Who Were the Fascists?* Bergen, 1980.

Phelps, R. H., '"Before Hitler Came"; Thule Society and Germanen Orden', *JMH*, 35 (1963).

Priebe, H., 'Zur Frage der Gestaltung und Grösse des zukünftigen bäuerlichen Familienbetriebes in Deutschland', *BuL*, 27 (1941).

— 'Wirtschaftsziele eines Umsiedlerhofes im Warthegau', *Neues Bauerntum* (1941).

— 'Betriebsgrösse und Betriebsgestaltung', in, ed. Wörmann, E., *Handbuch der Landwirtschaft*, 1952.

Puhle, H-J., 'Aspekte der Agrarpolitik im Organisierten Kapitalismus', in, ed. Wehler, H. U., *Sozialgeschichte Heute*, Göttingen, 1975.

Reischle, H., 'Die Entwicklung der Marktordnung des RNS in den Jahren 1935–6', *Jahrbuch der Nationalsozialistischen Ökonomie* (1937), 216–26.

Riecke, H.-J., 'Ernährung und Landwirtschaft im Kriege', in *Bilanz des Zweiten Weltkrieges*, Oldenburg, 1953.

Russell, Claire, 'Die Marktordnung im RNS-gewerbe', *ZGS* (1936).

Schoenbaum, D., 'Class and Status in the Third Reich', Oxford D.Phil. Thesis, 1964.

Schultz-Klinken, K-R., 'Preussische und deutsche Ostsiedlungspolitik von 1886-1945; ihre Zielvorstellungen, Entwicklungen und Ergebnisse', *ZAGAS*, 21 (1973), 198-215.

Schwarzweiler, H. K., 'Tractorisation of Agriculture; the Social History of a German Village', *Sociologia Ruralis*, 11 (1971), 127-39.

Schweitzer, A., 'Depression and War', *PSQ* (Sept. 1947), 32-37.

Searles, G. R., 'Eugenics and Politics in Britain in the 1930s', *Annals of Science*, 26, 1979.

Seraphim, H.-J., 'Neuschaffung deutschen Bauerntums', *ZGS*, 95 (1935), 145-54.

— 'Neuschaffung von Bauerntum und die Erzeugungsschlacht der deutschen Landwirtschaft', *ZGS* (1938), 625-51.

Smit, J. G., 'Neubildung deutschen Bauerntums: Innere Kolonisation im Dritten Reich, Fallstudien in Schleswig-Holstein', *Urbs et Regio*, vol 30, Kassel, 1983.

Stauffer, R. C., 'Haeckel, Darwin and Ecology', *Quarterly Review of Biology*, xxxii, 1957.

Strub, H., 'Die landwirtschaftlichen Genossenschaften im Reichsnährstand', in *Die Deutsche Genossenschaften den Gegenwart*, 3 (1937).

Timm, H., 'Zur Erbhofkreditfrage', *ZGS* (1938), 456-97.

Tracey, D., 'The Development of the National Socialist Party in Thuringia, 1923-30', *CEH*, 8 (1975), 23-49.

Vecoli, R., 'Sterilisation: A Profressive Measure', *Wisconsin Magazine of History*, 43 (1960), 190-202.

Verhey, K., 'Der Bauernstand und der Mythos von Blut und Boden, mit besonderer Berücksichtigung auf Niedersachsen', Göttingen, Diss., 1965.

Vlengels, N., 'Thünen als deutscher Sozialist', *Jahrbuch für Nat.-Ökonomie*, 153 (1941), 339-62.

Volin, L., 'The German Invasion and Russian Agriculture', *Russian Review* (1943).

Wachenheim, H., 'Hitler's Transfer of Population in Eastern Europe', *Foreign Affairs* (1942).

Wagemann, Ernest, ed., 'Deutsche Preispolitik und Weltwirtschaft', in *Vierteljahrshefte zur Wirtschaftsforschung*, 3 (1938-9), 333-51.

Wegner, B., 'Das Führerkorps der bewaffneten SS 1933-1945', Hamburg Diss. Phil., 1980.

Weiss, H., 'Die Umsiedlung der Deutschen aus Estland', *JBD* (1964).

Wheeler, Leonie, M., 'The SS and the Administration of Nazi Occupied Eastern Europe, 1939-1945', Oxford D. Phil. Thesis, 1981.

Wiseley, W. C., 'The German Settlement of the "Incorporated Territories" of the Wartheland and Danzig West-Prussia, 1939-45', London Ph.D. Thesis, 1955.

Ziche, J., 'Kritik an der deutschen Bauerntumsideologie', *Sociologia Ruralis*, 2 (1968), 105-41.

Index

NDG/140. Backe described 'honorary' SS membership as compulsory for all LBFs after 1933, *G.B.*, p. 36. Dr R. Proksch, ex-*Artamanen* leader, confirmed this question of cross-membership in correspondence, adding that 'underlings were not encouraged to join the SS'.

40. C. Snyder, *The History of the Death's Head Battalion 1933–45* (Princeton, 1977). Kummer and Willikens were both in fighting units in France, 1940–1.

41. Two weeks before the Night of the Long Knives, Darré wrote to Röhm, Head of the SA, enclosing copies of correspondence between Darré and the Chief of the SA leadership office, asking Röhm whether he approved of the SA letters, which contained 'a multitude of complaints on agricultural questions by the SA—unhappy with the RNS handling of settlement ...' and which asked Darré to discuss the complaints with the SA agricultural adviser. 24.5.34, BA R43II/207.

42. 20.12.35, BA NS2/137.

43. RFSS to Darré as *Bauernführer*, 30.4.36, BA NS2/50.

44. J. von Lang, *The Secretary*, p. 84. Backe personally warned Darré in 1938 that Himmler had his eye on the RMEL's Settlement Department, *G.B.*, pp. 40–1.

45. E.g. Darré to Himmler, 30.4.36 and 12.5.36, BA NL94 II/49.

46. 22.12.36, BA NS2/50.

47. Undated letter, Himmler to RuSHA, BA NS2/50.

48. Himmler to Dr Schmidt, RuSHA Settlement Office, 5.3.37, BA NS2/50. Kummer to SS Oberabschnitt Nord, 25.11.37, BA NS26/944. He added pointedly in his letter: 'In my report to the RFSS I will certainly make favourable references to this comradely help'.

49. Kummer to SS Oberabschnitt Nord, 25.11.37, BA NS26/944. Copies of the application form are in BA NS2/45. 3500 to 4500 RM was needed for a deposit on a farm and also livestock: cf. Farquharson, *Plough and Swastika*, p. 150, who mentions c. 15,000 RM as a minimum.

50. Letter from Kummer and Professor Emil Wörmann to Himmler, 27.11.37, BA NS26/944.

51. See Hofmann to RFSS on the problems involved with admitting *Kreisbauernführer* into the SS, 16.12.40, BA NS2/56.

52. Kummer to Himmler (my underlining), 27.11.37, BA NS26/944.

53. Pancke to Himmler, 10.8.38, BA NS2/54.

54. Circular from Darré as head of RuSHA, 14.12.37, BA NS2/163.

55. Undated memorandum, probably late 1937, concerning an SS

village settlement, which described procedures in a camp for selecting SS settlers, BA NS2/137.

56. SS OSF Klumm to Himmler's Chief Adjutant, 7.11.36, BA NS2/290.
57. NSDAP *Gauleitung* Swabia to NSDAP *Kreisleitung* Günzburg, asking the latter to support the SS recruitment drive, 4.2.39, BA NS2/290.
58. SA *Führung* circular, re Training of SS Agricultural cadres, 3.4.37, BA NS2/290.
59. Farquharson, *Plough & Swastika*, p. 153, estimates that 90 per cent of the *Verfügungstruppe* had been brought up on the land. See Wegner, Bernd, 'Das Führerkorps der Bewaffneten SS 1933–1945' (Diss. Phil., Hamburg Universität, 1980), for a contrary view.
60. The *Wehrmachtlehrgüter*. Draft proposals for *Ausbildung der SS Männer als Siedlungsanwärter*, 6.1.41, BA NS2/56.
61. Ideally, for the SS, he would be under 35, son of a peasant or agricultural labourer. See Farquharson *Plough and Swastika*, p. 159. Only thirteen per cent of settlers in 1938 were under 30. Seventy per cent were aged between 40 and 50.
62. See correspondence between Eggeling (the Keeper of the Seal for the RNS), and Goering, concerning Meinberg and Granzow's attempted putsch, 7.4.37, BA R16/2222, and Gies, Prof. H., 'Der Fall Meinberg', unpub. seminar paper, April 1981.
63. Pancke to Himmler, 10.8.38, BA NS2/54.
64. See E. Georg, *Die Wirtschaftlichen Unternehmungen der SS* (Stuttgart, 1963), pp. 133–4. See Darré to Himmler, 20.9.39, BA NS2/55.
65. Undated draft of letter, from von Gottberg to Darré in reply to Darré's letter of 20.9.39, BA NS2/55.
66. *RKFDV*, p. 242.
67. Pancke to Himmler suggested using the DAG for resettlement because the OKW had a good opinion of them, 20.11.39, BA NS2/60.
68. Correspondence and minutes of the DRV, 22.7.39 to 8.3.40, pp. 3 and 4, BA NS2/257.
69. May 1939, BA R43II/233a.
70. In October 1938 Darré wrote to Lammers after visiting the *Bayerische Ostmark* (Western Czechoslovakia) on the need for urgent aid measures for the mountain peasants there. BA R43II/202a. See also 7.10.38, BA NS10/36.
71. Darré corresponded secretly with Heinz Haushofer, then attached to the Austrian Ministry of Agriculture, who reported to Darré on the

degree of support existing there for Darréan ideas, 6.3.38, BA NS35/11.

72. Lammers to Darré, 8.5.39, BA R43II/223a.
73. Koehl, *RKFDV*, pp. 40–2 and 44–8.
74. The Sudetenland was annexed on 1.10.38 and Henlein appointed *Reichskommissar*. Rich, *Hitler's War Aims*, ii, 21. It was declared a *Reichsgau* on 30.10.38. The transfer of population agreement was signed on 20.11.38 and concerned Czechs and Germans who had moved to their present homes after 1910. Older residents had a right of option. According to Koehl, *RKFDV*, pp. 40–3, Germans were told to stay within the rump Czech state to provoke the Czechs, but neither German nor Czech minorities were disturbed. However, correspondence between a Czech co-operative and the Reich Chancellery in June 1939 mentions unfulfilled promises regarding their freedom of assembly and trade, and complains of 'harsh measures ... and bitterness among the local population'. BA R43II/221, and see Rich, *Hitler's War Aims*, i, 111–21.
75. Pancke to Himmler, 7.10.38, BA NS2/54.
76. BA R14/267.
77. Lammers File note, RK 211898, October 1938, BA R43II/208.
78. Rich, *Hitler's War Aims*, ii, 55.
79. *TWCN*, 14, p. 697.
80. Darré's diary, 26.11.39.
81. Darré's diary, 15.8.39. Czechoslovakia was occupied on 15.3.39. Rich, *Hitler's War Aims*, ii, 27.
82. 17.5.39, Darré's diary, Pancke's notes, 17.5.39, BA NS2/138.
83. Pancke's file notes, 3.2.39. A letter from Pancke to Kummer, 25.8.39, accused the latter of disloyalty and indiscipline, BA NS2/138.
84. Meyer, 'Lebensbericht', pp. 103–4. Himmler drove him to Cracow, while telling him to the plan to resettle the *Volksdeutsche* in the Warthegau, deporting the Poles to *Restpolen*, and warning Meyer that he would 'probably have difficulties with Darré'. Himmler drove rather slowly, complaining of careless driving by others, while his adjutant fed them bread and cognac from the back. Meyer was unable to convey his own excitement to Darré, who threw a pen at him on his return, with the remark 'Where matters of belief are concerned, you should go and see Rosenberg'. Ibid., p. 104. See, for Meyer's role as planning expert in Berlin in 1937–39, BA R43II/232, p. 220.
85. Darré's diary, 27.10.39.

86. See discussion, pp. 188–9 above and *TWCN*, xiv, 558–63.
87. Darré to Lammers, 29.9.39, BA R43II/613. The book was Wilhelm Rauber's *Bodenrecht als politisches Gestaltungsmittel: Gedanken zur Geschichte Irlands.*
88. Bergmann attacks Darré as a reactionary Imperialist over this issue, alleging that he 'legitimises his imperialism with national-Darwinist arguments', Bergmann, *Agrarromantik*, p. 308. The offending passage reads: 'The ethic of Blut und Boden gives us the moral right to take [back] as much Eastern land as is necessary to achieve harmony between the body of our people and geo–political space', *Stellung und Aufgabe.* As Darré habitually referred to East Prussia and the Baltic coast as the 'East' this passage does not seem to me to bear the weight placed on it by Bergmann, who seems to *invent* a passage in 'Innere Kolonisation' (Darré, 1926) to emphasise his point, pp. 308–9. See also Milan Hauner, cited p. 228, above.
89. Darré's diary, 19.1.38 and 11.9.38.
90. Darré to Himmler, 16.10.39, NDG/483.
91. Koehl, *RKFDV*, p. 28. Grundman, *Agrarpolitik*, pp. 73–4, describes how Darré's opposition to Himmler was not taken seriously by Hitler.
92. Darré's diary, 16.3.39.
93. Farquharson, *Plough and Swastika*, Ch. 11.
94. Darré's notes, NDG/182.
95. RMEL acknowledgement of decree to Schwerin-Krosigk, 16.10.39, NDG/483.
96. Darré to Lammers, 4.10.39, NDG/483.
97. Darré to Himmler, 16.10.39, NDG/483.
98. Himmler to Darré, 29.10.39, NDG/483.
99. See Koehl, *RKFDV*, Brandt, *Fortress Europe*, and Hans Merkel, communication to the author, 5.5.82.

NOTES TO CHAPTER SEVEN

1. Willikens, State Secretary, RMEL, to Himmler, 23.11.39, with copies to Goering, Lammers, Winkler, the OKW, the Finance Ministry, the Danzig civil administration, Gustav Behrens (*Reichsobmann*) and Dr Reischle of the RNS. My underlining, BA NS26/943.
2. See Brozsat, *NS Polenpolitik*, pp. 85–100, on *Nahplan* 1, 2 and 3.
3. The term 'slave labour' will be avoided in this discussion; it seems

to the writer that some form of barracks and/or curfew has to be in existence before workers can be termed 'slaves'. See discussions between RNS, Kaltenbrunner and Müller: RNS representatives forwarded requests from rural communities that barracks and curfews be imposed on foreign farm workers, 18.6.43, BA R16/162. See also investigation by RNS into complaints from foreign agricultural workers, 1944–5, BA R16/174.

4. SS SBF Brehm's Report, 24.10.39, BA NS2/60.
5. Brehm report. See also Rich, *Hitler's War Aims*, ii, 71. The border was not determined for several months, and was disputed among German officials until the war ended.
6. 26.11.39, BA R75/13. Poniatowski villages were the Polish version of the *Wehrbauern* settlement, which inspired Himmler. Border villages were fortified, and Germans expelled from them, during the 1920s. Koehl, *RKFDV*, p. 44.
7. 8,000 agricultural experts had been drafted by 1940, and 42,000 by mid-1941. Hans-J. Riecke, 'Der Raum in Osten', *DAP*, Oct. 1941. Riecke was head of Chefgruppe 1A Food and Agriculture in the Economic Command Staff East in 1941.
8. Brandt, *Fortress Europe*, pp. 36–8. Wheat down 21 per cent, rye down 38 per cent, barley down 21 per cent, oats down 30 per cent, potatoes down 12 per cent. See also Sering, *Agrarverfassung*, p. 179.
9. 'Beiträge zur Wehrstrukture Polens', *Reichsamt für wehrwirtschaftliche Planung*, July 1939, BA R24/788.
10. SS SBF Brehm, RuS Adviser to RFSS, 24.10.39, BA NS2/60.
11. Cited in Wheeler, 'The SS', p. 133. BDC Globocnik file. 'Bericht über Aufbau der SS und Polizeistützpunkte', 18.7.41.
12. See D. Warriner, *The Economics of Peasant Farming* (London 1939), pp. 21, 133.
13. A. Hohenstein, *Wartheländisches Tagebuch aus den Jahren 1941–2* (Stuttgart, 1961), pp. 53, 72.
14. Brehm report, *op. cit.* Underlining in original document, and see pp. 7–10, BA R24/788. Ironically, the wheat surplus area was in the Polish territory conceded to Russia. J. Gross, *Polish Society under German Occupation: the General Gouvernement 1939–44* (Princeton, 1979), p. 93; Brandt, *Fortress Europe*, p. 36.
15. Meyer, 'Lebensbericht', pp. 108–9. Intr., H. Heiber, 'Das Generalplan Ost' (docs.), *VjhfZg* 6 (1958), 289.
16. Himmler to SS leaders at Danzig, 24.10.39, BA NS2/60.
17. Backe, *G.B.*, p. 45. See also Kröger, Erhard, *Der Auszug aus der alten Heimat: Die Umsiedlung der Baltendeutschen* (Tübingen, 1967),

who makes a puzzled reference to Himmler's 'anachronistic perspective', p. 157.

18. Pancke to Himmler, 27.11.39, BA NS2/60.
19. SS GF Hildebrandt, 26.11.39, BA R75/13.
20. Unsigned report of conversation with Gauleiter Greiser, 12.2.41, BA R49/1/34.
21. Ibid.
22. Siegmund to RuS, 13.2.41, BA R49/1/34.
23. RFSS A/O. 24/1, 9.12.40, BA R49/4.
24. According to one report, 300,000 ha out of 3.4 m agriculturally usable ha. March 1941, and Report by Dr Schmidt of the *Aussenstelle Ost*, 8.5.41, BA R49/1/34.
25. Anon., 'Neubildung deutschen Bauerntums und Aufbauarbeiten in den Ostgauen', *Der Diplomlandwirt* (15.10.40).
26. Kummer (by this time drafted into the army) to Dr Horst Rechenbach, an ex-official of the RuSHA. 19.4.40, BA NS26/947.
27. SS SBF Kunzel, of the *Rasse-und Siedlungswesen Abt.*, Posen, 12.12.39, BA NS2/60.
28. Pancke to RFSS, referring to Hitler's *Mein Kampf* and a speech by Himmler on 13.12.39, saying 'I would like to locate a blonde province here', 28.12.39, BA NS2/60.
29. Hofmann to RFSS, 17.840, BA NS2/50. Himmler rejected the pamphlet as being 'too theoretical for ordinary people'.
30. Kummer to the Danzig-West Prussia Settlement Society, 6.12.40, BA NS26/943.
31. Schöpke, Prof. K., *Deutsche Ostsiedlung* (Berlin, 1943), pp. 6–7, 32–3, 44, 54–5. Idem, *Der Ruf der Erde: Deutsche Siedlung in Vergangenheit & Gegenwart* (Leipzig and Berlin, 1935).
32. Auhagen, Dr. The report of 1940 was drawn up for the RKFDV. 'Die Ansiedlung deutscher Bauern in den eingegliederten Ostgebieten', BA R49/20. Auhagen listed the various exchanges of population which had taken place since the Versailles Treaty. In 1937, the same exchange proposal had been made to compensate Poles with Russian land, thus 'forging a German–Polish alliance'. Zimmermann to Hauptmann Wiedemann, BA NS10/105.
33. 'Proposals for the Agricultural Settlement of the new German areas', by G. Blohm et al., quoted in Hartmann, Peter R., 'Die annexionistische Agrarsiedlungspolitik in den sogenannten "Eingegliederten Ostgebieten" (Reichsgau Danzig-West Preussen, Reichsgau Wartheland, Regierungsbezirk Zichenau)', Appendix (Rostock, Diss. 1969).